D1409693

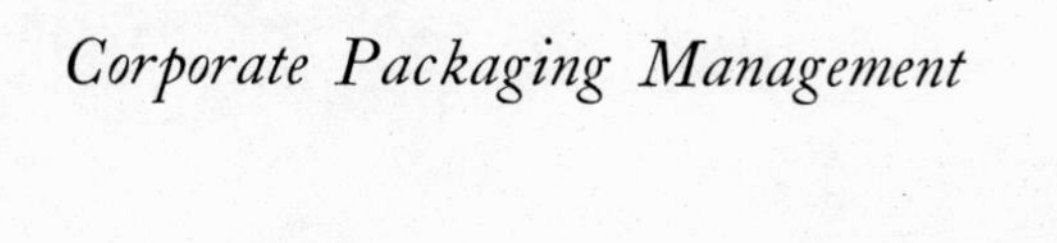

Corporate Packaging Management

About the Author

C. WAYNE BARLOW is manager, package material purchases, the Quaker Oats Company. Prior to joining Quaker Oats Mr. Barlow held various jobs wth Sears, Roebuck & Company, including buyer, general purchasing department; assistant general merchandise controller; and senior buyer, merchandise packaging. He is past president of the Mail Order Association of America and past vice president of the Retail Merchants Association. In addition to having authored several publications, Mr. Barlow's activities include lecturing at the Universities of Michigan and Rochester. He is a graduate of the University of Chicago Law School.

Corporate Packaging Management

C. Wayne Barlow

American Management Association, Inc.

This book has been distributed without charge to AMA members enrolled in the Packaging Division.

Standard book number: 8144–2121–0
Library of Congress catalog card number: 70–80889

To

Ruth, my wife, for her understanding,
Karen, my daughter, for her confidence, and
Bruce, my son, for his faith

Contents

Chapter I

Packaging and the Marketplace

The advent of self-service merchandising has highlighted packaging as a vital marketing tool. However, this was not always the case. Packaging as a recognizable, specific form is a product of the last 200 years. Before the 1800's the only concept of packaging was that of a container, such as a wine cask, for holding and transporting a product. In those early days there was practically no thought given to preservation from spoilage by use of a package, nor was protection of the product a matter of prime consideration. Early records indicate a minute use of grocery bags in the 17th century, but only on a handmade basis; not until the 19th century did practical bag-making machinery become commercially available.

The 19th century saw many other breakthroughs in the development of distinct packaging types. The metal can was patented around 1810; the Owens bottle-making machine became a reality in 1899. This period evidenced a change in packaging concept from mere containment of a product to the more modern idea of packaging—a fundamental shift that led to such outgrowths as product identification, sanitary packaging, product preservation, and the utilization of the package as a method of mass communication.

Once the concept shift took hold, the logical next step was a dramatic alteration in marketing strategy to incorporate the growth of brand identification and point-of-sale display. The transition from a static, rural society to a highly mobile, urban, industrial society and the consequent changes in living styles further spurred development of the market-oriented package.

Relationship of Packaging to Marketing Concepts

There is a definite complementary relationship between packaging and price, product, and promotion. These marketing variables are used in different combinations in planning for successful marketing, and in any given situation there must be a balance between packaging and each of these variables if there is to be a maximization of profit. The role of packaging in relation to each variable must be studied in the light of corporate marketing objectives; normally, a good marketing program equitably balances all the variables related to packaging in the same way that the basic elements of packaging costs—that is, materials, design, and services—are balanced.

Price. Because packaging offers the customer a range of possible price choices, based upon a variety of package sizes, it is a key cost factor in any product and can often be the determining factor in the product's success or failure.

Product. Packaging is a part of the product and, if properly utilized, can do much to enhance the appeal and probably the ultimate sale of the product to the customer. The package helps the product survive the competition—it provides esthetic appeal to attract and "sell" the customer and can create the proper friendly image with him to improve the quality concept associated with brand identification. In fact, in many industries it is virtually impossible to conceive of the product without the package.

- *Promotion.* It is here that packaging has perhaps its greatest impact. It is a *key* element in promotion, even though packaging costs are often considered as manufacturing rather than as promotional costs. Nevertheless, a package that does not do the job in its main functional areas and does not get the manufacturer's message across is a waste of money and virtually destroys the effect of promotional expenditures in other areas.

Usually, there are only so many dollars that can be allotted to promotion, and the trick is to balance expenditures in packaging, advertising media, salesmen, and the like to produce the maximum sales dollar. It is normally the intent, in approaching marketing, to define its objective as the expenditure of funds to gain the maximum sale of a given product at a given price. In this relationship, the marketing obligation of packaging is to influence product appeal in the marketplace.

Since the package is the keystone of distribution, manufacturing, marketing, and consumer response, it is one of the most potent forces in the economy. An industrial society cannot grow or reap its normal rewards without packaging to house the multitude of products that form its economic backbone. Accordingly, the development, design, and manufacture of a package present a complex process that requires well-trained professionals in every area.

The Corporate Packaging Team

The complexity of packaging management becomes instantly apparent when the list of departments involved is examined. There is no order of importance in which these contributing areas should be reviewed, except that there are three departments that actively participate in all packaging projects as a major activity while the others may or may not be involved in all projects. Therefore, in making a classification for illustration purposes, the involved departments will be classified as "primary" or "secondary." This classification relates only to their relative participation in *all* packaging projects and has nothing to do with their decision-making capabilities in the corporate hierarchy. The primary departments (often referred to as the "big three" of packaging) will be examined first.

- *Packaging procurement.* This is probably one of the most powerful areas in the corporation as far as packaging decisions are concerned, because procurement, equipped as it is with the best in technical talent and management orientation, is involved in almost every decision made during the development of a package.

In most companies that are "heavy" on packaging, the procurement people act as consultants, advisers, and often father-confessors. Their technical know-how is invaluable to their associates in packaging; it can eliminate errors and potential trouble spots before they occur. Procurement aids the R&D people in constructing the physical package and setting the specifications, it assists the design area on a consulting basis to help the developing team create a package that will be both economical to manufacture and yet within the technical limitations of the converters involved, and it makes available a wide range of converter facilities for use by both R&D and design. In many companies, procurement also acts as a clearinghouse for information, a coordinating center for current projects, and a planning center for potential packages.

- *Packaging research and development.* This most important area is greatly responsible for developing the physical shape of the package and for preparing the specifications for the procurement area. It is highly involved with design, because the physical shape of the package and the material chosen have a great effect on what the designer can do. It deals with procurement in developing the material specifications so that economics may be considered. Another of its responsibilities, which is often overlooked, is the testing of materials or packages for their effectiveness in doing the job they were designed to do. Such matters as breakage, odors, spoilage, opening devices, residual taste, and a host of others are the normal confines of this department. Its interests extend throughout the corporation, touching practically every area concerned with packaging.

Packaging design. The third part of the "triumvirate" is the so-called creative area that is concerned with design concepts and with the impact of the package in the marketplace. This is the department where the design and art production people are generally found, and although there is no contesting the effect a talented design director can have on the image of the corporation in marketing, design has another important responsibility—to act as liaison for product management, product development, packaging R&D, and packaging procurement. Packaging design bears a great communications burden; hence coordination is one of its most difficult problems.

Although these big three departments carry forward the difficult task of transforming a gleam in a manager's eye into a package that is a potent selling tool in the market, the secondary departments also play a role in package development. Despite the fact that some of them are involved in almost all packaging developments, they ordinarily function as advisory agencies.

Marketing (or advertising). Marketing's association with a package extends throughout the package's various stages of development. Initially, it is responsible for preparing the marketing base or "selling concept" for the packaging of a given product, and until it determines the concept and provides initial guidance and basic direction, the primary departments cannot make any definitive effort. Marketing prepares the market plan and time schedules as soon as the new product is approved so that all elements of the corporation will know what is expected of them, and it must also arrange for and conduct all test marketing and consumer-reaction tests on both product and package. This department remains involved with the package by approving design sketches, final design, initial press proofs, and color standards on the production run. It keeps close ties with production planning and product development as the product is conceived, developed, manufactured, and packaged. Contact with the sales force is also essential for promotion, test marketing, and national roll-out for general distribution. Marketing and packaging are inseparable in a packaging-oriented organization, since without the proper package no product will be on the shelves for sale.

Product development. This area's involvement with a package originates in the overall market plan for the corporation. Usually, such plans are made three to five years in advance on a lead-time schedule carefully designed to attack new markets, balance sales in certain areas, and challenge competitors' positions in relation to the corporation's market posture. Generally, the need for product *X* is determined, a product form is achieved and approved, a new close-in marketing plan is developed, and the package orientation begins to emerge. But as a rule product development keeps out of packaging development, since this is an area outside of its particular discipline. On the rare occasions when incursions into the packaging area do occur, top management should step in and discourage such unneeded participation. There is enough confusion in the average corporation's packaging development effort without adding the well-meaning interjections of peripheral departments.

Legal. The company's attorneys check and register trade names and edit package copy for compliance with state and Federal regulations.

Engineering. This group becomes involved at the very beginning of a project in order to provide manufacturing equipment when needed. Its major contact areas include production planning, machinery purchase, and product development, but it also has contact with the packaging design and development groups, since it must provide the layout detail for the package after its physical form has been determined. The engineering function is most important in the corporation's planning processes because it must prepare the plans for all production facilities, including equipment, that require capital expenditures of any consequence. A mistake here can be most expensive in terms of time and money.

Machinery purchase. This activity is a vital part of the picture since it is responsibile for procuring the necessary equipment, and very often it must also expedite and oversee the entire construction of a new facility. Its major areas of contact are engineering, production planning, and product development. The legal department assists in the preparation of all major contracts for equipment and facilities.

Traffic. This department supplies the expertise for arranging and evaluating test shipments and for contracting special means of transport. Intelligent use of this group can ease many problems for both packaging development and procurement.

The Corporate Packaging Dilemma

In an increasingly competitive marketplace, the impact of the package is becoming more important each day. Unfortunately, however, few companies handle the management of packaging in a manner consistent with the management skill apparent in other areas. In view of the competitive value of a good package and, therefore, of its importance to the corporation's success, it is unusual that it is accorded so small a measure of management sophistication. The fact is that packaging management is not keeping pace with the increasing skills of the packaging designers and producers and is failing to meet the needs of the corporation. Changes in the packaging industry, such as new materials and advanced production techniques, are making packaging decisions more complex, more time-consuming, and considerably more hazardous if taken without proper background information. The "right" decision is a most difficult problem; therefore, today's company needs a management orientation that permits the corporation to capitalize on the wide range of skills and capabilities that must be utilized to make an effective, timely, and economical packaging effort a functional reality.

It would be foolish to say that package-oriented corporations are not concerned with this management desert, but it is true that the effort has not been aggressively pursued. Packaging as an industry is big business. Estimates place it, all aspects included, in the $26 to $30 billion class. Since packaging costs represent a large portion of the final cost of a given product, top management should be alert to the need for an adequate, active, and functional management response. This is not generally the case.

One of the major reasons why packaging presents such a management problem is that it is not amenable to the accepted management rules. It is a complex field, filled with activities that defy the traditional approaches to the placing and structuring of responsibility and authority. Proper management of the packaging effort requires a peculiar blending of organizational structure, responsive personalities, and a tremendous amount of marketing orientation. The marketing orientation factor in the packaging effort is a must, but too often the major areas that concentrate on the technical aspects of the package, such as physical development, design, and purchasing, are left out of the market planning and discussion. This is a fatal mistake, since the technical concepts have a distinct relationship to the image that the marketing group wants the package to convey. Suggestions and help from the technical people, given early in the planning stage, can more easily assure the desired package. The alternative is to settle for a package that does not quite live up to expectations because of technical limitations of the production equipment. Generally, the final package is the result of many compromises, but this fact does not constitute a weakness in the packaging effort. Rather, it indicates a continual "meeting of the minds" between the marketing people's desires and the technical people's production limitations—limitations that marketing people, in general, are not aware of. The problem revolves around how to mesh these areas of interest.

Although there is some divergence of opinion on the best way to tackle the packaging management problem, there are some basic considerations that are useful, particularly to a consumer-oriented company. The first of these is that there is no one, perfect, foolproof method of handling the packaging effort. An approach that works for one company may not be at all acceptable for another because of the wide variances in management concepts, reporting structures, and personalities involved.

Another consideration is that, as has been noted, packaging management usually does not lend itself to traditional management techniques. Essentially, it is a creative business with a variety of potential solutions. Such a bewildering array of potentiality must be subdued by management approaches that border on the inspirational. The need for this enlightened approach is evidenced by the number of differing opinions that must be compromised, directed, and even submitted to mediation before any packaging effort is completed and by the problem of controlling the often strong personalities involved. Today's management must be willing to consider and try new techniques, and the more ingenious the management approach the more successful the packaging effort. An understanding of what motivates people, of how to control personal energies, and of the general field of psychology is essential. A successful packaging operation is a blending of talents, of personalities, and of systems to allow free expression and utilization of individual talents—all clearly directed toward a corporate goal and functioning within a framework of authority and flexibility.

The final decision on a package is determined by the skill, judgment, and knowledge of more diverse areas and more diverse personalities than probably any other corporate decision. This is further complicated by the fact that these diverse information and talent sources are usually located in different corporate command structures, thereby leading to almost inevitable power struggles between department heads. Add to this the individual's need to better his own lot while doing a job for the company and the web becomes more entangled.

Packaging management as a line function. The proponents of assigning packaging responsibility to the "line" category have taken an almost military tack in attempting to simplify the problem. It is true that giving control of the packaging function to a line department tidies up the organization chart, but it is not the most effective approach. If a division for handling all packaging problems is created, many technical packaging specialists must be brought under one administrative control. This may be neat and orderly, but it is not really practical because it overlooks or preempts the participation of other vital factors in the packaging world—the marketing people, packaging R&D, and product development. There is also the question of whether it is profitable to gather all this talent into one area.

The main problem with the line approach, however, is one of operating philosophy: Does the head of this "packaging department" have the power to override the judgments of the heads of other interested areas? To have his department function effectively as a line operation, this must be the case; however, in reality this cannot be done within the modern corporation's packaging concept. There are too many interdepartmental concerns that affect the performance and profits of other managers and of the overall corporation. An additional problem with the line concept is the level of authority to be assigned. There is such a mixture of levels involved in making packaging decisions that any consolidation of control would probably demand assigning overall responsibility to a very high level. Here again, the possibility of one manager's veto affecting the performance of another area of the corporation becomes even more pronounced and an even greater potential area of conflict.

On final analysis, it would appear that, from a practical operating viewpoint, packaging should not be set up as a line function in most corporations. In relatively small companies where the volume of work is not so heavy or the packaging emphasis is on mere protection or containment of the product, it is possible that packaging can work effectively as a line operation. But in such situations the control is generally located in the purchasing department.

Packaging management as a staff function. The advocates of this philosophy feel that there should be a headquarters staff established to act as a clearinghouse, a control and information center, and a general command post. On paper, the concept looks good—until it is dissected to reveal the real weaknesses. The first problem stems from the fact that a staff function, if properly established, is an advisory operation. It does not have a decision-making ability but merely gathers data, advises, and assists operating executives who have a profit responsibility.

The second area of difficulty arises because, to function at a high level with confidence, the staff must be fully equipped with a wide range of expertise. This means that the staff begins to infringe upon the operations of the various segments that are doing the actual work. What results, in fact, is a virtual duplication of effort in the development of packaging, which, in turn, creates further industrywide problems owing to the great shortage of all types of skills in today's packaging industry. There is also a psychological problem. Since the specialized line departments, such as the "big three," try to equip themselves with the best talent available, there is always a tendency to resent the intrusion of a staff man who is slipped into an intermediate slot between line and top management. The line generally feels that there is an unnecessary hitch in communications, and the unfortunate staff man is regarded as a natural enemy. This subconscious resentment—and it does exist—is usually readily apparent to the staff man.

The lot of a staff man is, at best, dismal; the job requires a very special kind of individual. He must be a master of discretion, have a tremendous talent for persuasion, and have an innate sense of timing—that is, he must know when to put on the pressure and when not to. Above all, he must always remember that he does not have a command function. This last requirement presents the greatest problem, particularly while the staff man is learning the job. There is a tendency for the inexperienced to exert pressure to get the job done; this breeds resentments that take time to cure, if they are ever really healed.

Even if all these disadvantages could be tolerated, the problem of the staff's dubious authority would remain, particularly where there were sharp divisions of opinion and where other departments' records were concerned. If it were established, it is doubtful that the packaging management staff would end up being more than just a powerless debating society. There are many places in the corporation where a staff function works well, but packaging is not one of them.

Who really has responsibility for packaging? It is evident that, to come up with the kind of packaging that will do the best job for the company, there must be a multitude of decisions made at all levels. Some of these, such as decisions on marketing strategy, profits, advertising policy, and corporate growth objectives, must be made at the highest levels of the corporation, because they involve such key matters as the corporate posture in the industry and in the marketplace. Obviously, such decisions cannot be left to lower echelons, to a staff function, or to an individual heading the "line" packaging department. This leads to the proposition that packaging for a multiproduct company or a multidivision operation cannot and should not be decentralized. Decentralization of the packaging effort, epitomized by development, design, and procurement, culminates in a "bag of troubles" that no corporation wants. The overall marketing and sales strategy of a corporation is not a delegated decision, and since the packaging effort is a necessary adjunct, it also prospers best under centralized control. Sound and reasonable packaging decisions, focused on corporate image, design consistency, and constant quality, are essential for proper utilization of packaging as a sales tool. Therefore, packaging decision-making authority must be lodged at a high enough general management level to permit disputes to be settled and decisions to be made on the basis of the greatest good for the overall corporation. It is only with this type of corporate-minded thinking that a company can hope to extract the last bit of yield from the packaging dollar.

Packaging committee. Even if the concept of final, high-level packaging responsibility is accepted, the problem of developing an organization that will provide top executives with all the necessary background information for making wise decisions still exists. More and more organizations are trying the committee approach or are appointing packaging coordinators to do the necessary legwork. But corporate history has proved the ineffectiveness of the committee system of problem solving. In the packaging area, it is generally agreed that a packaging committee, established on a usual committee pattern, is a relic of the past. It does not work effectively in most companies, because it does not accomplish the primary missions, which are: (1) to insure that all of the innovative capabilities inside and outside of the company are brought to bear on the problem, and (2) to provide any assistance needed over and beyond the skills available. Other disadvantages are that committees are normally too big, composed of the wrong people needed to get the job done, harassed by changing personnel, and difficult to convene with a full working complement.

It is true, however, that in certain corporations the committee approach works very well. But a further look at these operations reveals a key factor in their success—that is, the committee members regard their committee obligations as major responsibilities that have top priority on their minds and time. Their seriousness of purpose and dedication to the committee concept are bound to make the system work. Such effective packaging committees are the exception rather than the rule, however.

Packaging coordinator. Because they doubt the effectiveness of committees, many corporations turn to the idea of a packaging coordinator as the answer. This is swinging the pendulum the other way, and although the concept looks good on paper, in actuality it has problems. To be effective, a packaging coordinator must be a man of many superior talents. Ideally, he should be soundly based in the fundamentals of development, design, purchasing, and marketing—a big order to begin with. Add to this a few other requirements: A coordinator must have a broad management orientation so that he can function at almost any level; he must have the ability to see the corporate picture in its entirety and yet appreciate the problems of the packaging people at the same time; he must be at an organizational level where he has recognized and validated decision-making authority; he must be able to handle people; and he must be a common meeting point or liaison between the company and suppliers of packages, equipment, and services. If such a man could be found to fill the position, if he were properly backed by top management, and if there were excellent cooperation and collaboration by all the main contributors to the packaging area, the packaging coordinator approach could work—and probably very effectively. But the lack of available talent would seem to preclude this approach.

Packaging team. After studying the pitfalls of the various one-tack possibilities, it appears that a middle-of-the-road approach might be the easiest to implement. With all the emphasis today on the "corporate team," perhaps a "packaging team" is the logical answer. The packaging team is, in essence, a committee, but with considerable differences. A team is based on function and cooperation. It is a group, whether it is an infantry firing team or a corporate packaging team, that avoids the pitfalls of the committee system because it is small, stocked with talent, and highly functional, and that does not have the problems of the coordinator, because it consists of the heads of development, design, purchasing, and marketing. The key decision-making elements are there—viewpoints can be expressed, evaluated, and consolidated, and the team can make decisions with a minimum of effort and delay. In addition, it is highly flexible and can be supplemented with experts from engineering, production, advertising, market research, and the like when needed. The very act of asking for help from interested groups tends to eliminate corporate jealousies and bring out the best advice available. But the team's largest plus is its ability to define the problem and make a decision. If it is properly organized, properly oriented philosophically, and properly backed by top management, the team is a functional force capable of handling the packaging job.

One must always reckon with the diverse personalities that abound in business and with *sub rosa* obstructionist tactics; but the face-to-face confrontation of the team approach to a problem quickly pinpoints the "nondoers" and those afflicted with "status quo-itis." If obstructionists are present and are hampering the packaging activity, team pressure usually can resolve the problem; but if it cannot, the problem is at least exposed for top management action. There is no place to hide when a team is functioning properly. Those who will not or cannot carry their full share of responsibility are instantly apparent, and corrective action can be taken. Although any approach has potential trouble spots, the team concept seems to offer the best hope in the current corporate setup.

Chapter II

Packaging Research and Development

PACKAGING'S research and development division is one of the key departments in the overall organization. Traditionally, it encompasses the more technical aspects of packaging and is involved with a wide range of other interest areas. If a general definition of the basic function were devised, it might read as follows: The basic function of packaging R&D is to plan, direct, and control packaging development activities throughout the company in the areas of both materials and mechanical equipment and to perform an interdepartmental coordinating function with regard to the corporate packaging effort. The extent of this group's workload is limited only by the number of personnel available, their backgrounds and capabilities, and the needs of the corporation. It is essentially a package engineering and testing operation geared to the interests of the corporation.

Planning Responsibility

The R&D group is responsible for developing and recommending long-range packaging plans, cost-reduction programs, broad-based packaging material and equipment programs, and programs to improve existing packages and equipment. The necessity for long-range planning in the business climate today is such that to fail to plan is to court economic disaster. Opinions vary, but the most workable approach seems to be a three-year plan stating general objectives in sequential order, coupled with a one-year plan of specific missions to be accomplished on a given schedule. In preparing a long-range plan for packaging development, the development manager must first draw on the objectives and changes planned by others—for example:

- What is the new-product introduction schedule?
- Has market analysis uncovered new outlets that may present different problems?
- What are the machinery development trends in packaging? Illustrative of the principle involved is the growing tendency of packaging suppliers to supply not only a package but the packaging machinery.
- Are there changes in the purchasing group's strategy? If purchasing pursues cost savings aggressively, the development plan must provide adequate manpower for its program. In most packaging-oriented companies where R&D tests packaging materials, specification changes cannot be made before testing is done. Cost-savings programs usually depend on specification changes that allow substitution of a less expensive material; hence, lack of R&D manpower inhibits testing and hinders packaging cost savings.
- What manpower requirements are indicated by the corporate growth plan? It is essential that the manpower plan be accurately developed and implemented on schedule. If this is not done, the growth of departments such as marketing and purchasing, which depend on package

> development, could well be hindered. Lack of manpower is no excuse when the progress of the corporation is being slowed down by failure to provide the services required.

Rarely does the first packaging proposal become the ultimate package for a particular product. All too often, development people tend to delay new-product introductions by striving for the "perfect package"—a practice that can be harmful to the corporation. The preferable approach is to keep an "open file" attitude—to set a workable schedule, adhere to it, develop a reasonable variety of packages, conduct an honest evaluation, choose the package that appears to fulfill most of the assigned objectives, put the package on a systematized review schedule, and actively work to improve its physical appearance. This is generally the attitude of design people, who realize the need to update and freshen graphics. The same general principle of improvement applies to package development.

With some modifications the evaluation process for packaging machinery is essentially the same as that for packaging development. The money and the lead time involved in changing equipment call for careful consideration and analysis. Demands vary by industry, but often the existence of expensive machinery limits the packaging group's ingenuity in developing more exotic packaging. In analyzing a project that involves replacement or addition of equipment, the development group must consider several fundamental factors:

- The basic cost of the equipment and its effect on the overall corporate plan.
- The effect of the machine and its output on product costs as compared with the corporate investment required.
- The feasibility of the package requiring the machine in terms of the package's potential impact in the marketplace (determined by consultation with marketing). If, for example, there is little real evidence that the new, exotic package will give the corporation a definite market

advantage, it should be reviewed with a wary eye. The package must pay its way at all times. To a great extent, this evaluation is the least scientific of all.

There is probably more confusion in planning than in any other activity; and yet the better the planning, the more efficient and orderly the effort. This is why planning is especially vital to a group such as packaging research and development, which can directly affect such areas as new-product introduction, packaging cost, and capital expenditure. Planning is the key to timely accomplishment of tasks and is an essential ingredient of corporate growth.

Problem Solving

R&D has a definite responsibility to assist almost every department in solving technical problems. This is particularly true where it has a heavy responsibility for industrial packages, such as shipping containers, designed to protect the item until received by the customer.

In most cases it is not necessary to make a huge investment in container-testing equipment because a large corporation's R&D facilities are usually equipped to perform all the standard tests, such as compression and vibration, to simulate shipping and warehousing conditions. And most large corrugated-container companies have testing and research facilities in metropolitan areas. It is perfectly ethical to utilize these facilities, particularly when the container company is supplying a portion of a corporation's needs. Using source facilities is generally not a complicated procedure, but it is advisable to consult with purchasing on the suppliers and then try to stay with those selected as a hedge against problems arising from unfamiliarity with specific company requirements. In many respects, the R&D group functions as the right arm of purchasing, and, when the two can function well together (often quite a trick), the corporation reaps great benefits.

Packaging R&D is often called on to improve a retail package that is causing trouble. The major reasons for such a request generally are:

- Pilferage at the retail level.
- Insufficient protection of the product, causing returns and customer dissatisfaction.
- Shipping problems stemming from Rule 41 restrictions. (This rule, which clearly outlines the package-strength requirements for given categories of goods, is part of the Standard Freight Classification restrictions established by the railroads.)
- A merchandising desire for a new package image to give the item a fresh look on the counter.

The food industry offers some particularly difficult areas for problem solving—odors and taste transference, for example. Odors present almost unbelievable problems, particularly in the cereal and snack fields. Many times, the physical components of a product have a particular affinity for attracting and retaining odors or tastes.

The in-package premium, a favorite merchandising gimmick of cereal producers, often leads to some very definite odor and taste problems. These premiums range from printed inserts to plastic and wooden toys, stamp sets, coloring books, and a host of other items, all with odor and contamination possibilities requiring special attention. All premium items are usually prepackaged in a plastic film to decrease the potential dangers. Tests for leaky or broken seals are very stringent, with some companies requiring that 95 percent or better of the seals be perfect. Printed inserts present the most problems because the inks must be approved by the Food and Drug Administration as nontoxic, and the inserts' potential for odor and taste transference is great. All ink must have a solvent carrier, so there will always be an odor of sorts; the so-called good inks have very little odor but very dull colors. The best way to prevent trouble is to have base stocks tested and approved for various types of cereals, then allow at least three weeks for aeration before use. All packaging materials must be tested and cleared for odor transference.

R&D's problem-solving responsibility often has a bonus for the company. It is not unusual for one of its problem-solving tasks to blossom into a full-scale research effort, leading to unexplored fields with great packaging potential.

New-Package Development

This is probably the most important and exciting realm that R&D works in. The challenge is far-reaching, and, when a project is completed, the satisfactions are great. The topic will be approached from both a general merchandising and a food industry viewpoint. The development of a package in a manufacturing company generally starts with a request from the engineering department. Since protection of the product is paramount, the complications are usually fewer than in other packaging efforts. This in no way demeans protective packaging but merely indicates that there may not be as many problems because there are fewer kinds of materials to screen and evaluate.

General merchandising new-package development. The new-package request usually originates with the merchandising buyer and merchandising staff. Normally, the packaging group then meets with the buyer and with the art director. The item requiring the package is displayed and discussed with special interest given to:

- Package-cost limitations, if any.
- Selling concept—that is, the market the product is aimed at.
- Image desired (super quality, utilitarian, masculine or feminine, fun or recreation).
- Display concept.
- Other pertinent data that will give the package development people a feel for the item.
- Plant packaging techniques and equipment to be used.

The packaging project is then assigned to an individual for

completion. Normally, the engineer or designer responsible has a basic feeling for the concept he wants to pursue. Next, purchasing calls in packaging sources to compete for the job; they provide even more ideas and samples. At this juncture, care must be taken to insure that all pertinent data and limitations on equipment are made known to those on the project. When all the samples have been brought in and tentative costs assembled, the package development group again meets with the merchandisers. At this meeting the final direction is set, design gets a shape to work with, and the project is well under way. Then the final sample is approved, the specifications are written for purchasing, and the design group prepares the graphics. Purchasing takes over when it receives the final artwork.

Food industry new-package development. The food industry follows the same general procedure, but with more detailed testing. The initial packaging request generally comes from marketing to R&D by way of the product development group, which is responsible for scheduling and forwarding the development of a new product. Product development's schedules incorporate the myriad details involved in a test marketing and, eventually, a national marketing arrangement.

After the project is activated, R&D consults with purchasing on packaging sources to use in the development phase. The importance of a source to the development people cannot be overemphasized because the source usually develops most of the physical package shape while R&D does the long-term testing and materials and equipment evaluation studies. When several seemingly acceptable packages, which may consist of different types of packaging materials, are ready, they must be tested—a time-consuming procedure because of the extensive examinations that must be performed on each package to see how it fulfills specific requirements. One of the major requirements is protection of the product, and the package is subjected to such tests as the following:

- Degree of product breakage. The recommended package is tested in a preliminary fashion to determine what breakage of the product occurs. Cereals are, for the most part, extremely fragile and susceptible to excessive breakage from extreme vibration, "loose fill," or crushed cartons resulting from unusual stacking methods. R&D must thoroughly test each proposal before making a recommendation.

- Shelf-life requirements. Selecting materials to provide the proper shelf life is a most difficult procedure, and it begins once marketing has specified the shelf-life requirement. Most packages have an inner lining of glassine or foil; some require a pouch-paper combination. The package material must be machinable at economical speeds, be easily workable, have consistent quality, and provide the protection needed. In tackling the materials-selection problem, the technical services of the basic materials sources—that is, papermills, extruders, and laminators—are most valuable. To be safe, R&D must conduct tests over the time period specified for the shelf life. Accelerated tests are possible in short-time situations, but for R&D to be secure in its recommendation, the actual specified time period and closely simulated conditions must be used. Simulation becomes quite complicated if humidity affects a product. In this case, the material specification is designed for certain climate zones—usually, a form of protected foil for high humidity areas and other laminates for areas of lesser humidity.
- Palatability. This area of shelf life and palatability can get quite complicated because various products have characteristics that require different approaches. Generally, palatability refers to taste retention, avoidance of staleness, and the absence of perceptible transference of odors or tastes from foreign objects, such as premiums or inserts. Some products must breathe, some need various preservative gases, and others must be fully protected against moisture and other vapor transmission.

Product protection is only one of the requirements that the package must fulfill. There is also testing of various packaging shapes suitable to the current equipment, new equipment (if new equipment is required), newly developed or heavily adapted equipment, shipping and warehousing facilities, and advertised weights (by performing filling tests).

Material Specifications

Once the tests are completed and the optimum package combination is selected, the specifications must be written, and it is essential that they be clear, thorough, and workable. The specifications then go to purchasing, which must buy the materials. Purchasing should have the option to reject the specifications if they are:

- Unrealistic in that R&D attempts to eliminate all risk by a massive overspecification of material. It should be pointed out that this is often the case when R&D is pressured for results on a short schedule and when it does not have time to complete adequate testing. The position of R&D in this situation is most unenviable, and it can be forgiven for seeking an escape route.
- Neither economically feasible nor readily obtainable. There is always a fascination for a new material. Often, R&D experiments with pilot production of a new material and specifies it before it is commercially available. If this happens, chaos results, but it invariably leads to a closer cooperation with purchasing. If the materials specified are such that they wreak havoc on the package-cost limitations, the project must be re-evaluated and more economical alternatives selected. If this situation occurs too often, the result is usually a radical shakeup of the R&D organization.
- Unnecessarily inhibiting to purchasing. Specifications should never call for a given supplier's brand of material unless that is the only supplier of the item. Many times, R&D's specifications inhibit the ability of purchasing to

indulge in competitive bidding. In this situation, purchasing should reject the specification outright and demand a general rather than a confining specification. For example, if a carton board is needed, purchasing should make sure that it meets all the required specifications, but R&D should not request a particular board under a trade name. R&D's tendency to specify by trade name or supplier can be cured with some understanding and education and by the development of a basic materials orientation.

While the specifications are being written, a machinery evaluation study should be conducted by the engineering department and the machinery buyers in the purchasing department. It is vital, particularly with a new application, to thoroughly explore any possible machine complications early in the game. This aspect of packaging development is a major part of R&D's coordinating responsibility. Failure to perform this function leads to many unnecessary difficulties and measurably affects the company's product introduction schedules.

Relationship Between R&D and Purchasing

Communication is one major problem that seems to plague all large corporations. Failure to communicate leads to misunderstandings, missed schedules, financial losses, and, many times, to outright battles between both individuals and departments. There does not appear to be any pat answer to the problem; memorandums sometimes violate security, and the list of recipients grows constantly. Meetings become too numerous to be effective because of the difficulties of finding a suitable time for all concerned. The communication and interrelationships of the various key departments present a classic corporate dilemma, and R&D's task in this area is most important and most difficult.

In companies where R&D does not report to the head of purchasing, its most difficult relationship is with that department.

The reasons for this are basic—they center around the utilization of outside packaging sources for completion of the development work. R&D usually has strong feelings about its need to delve, without restraint or limitation, into areas it wishes to explore. Its general attitude is that it should be able to call in anyone it wishes for development work. Needless to say, the average purchasing man or group takes a somewhat dim view of this idea. There is no arguing the fact that R&D's members must have intellectual freedom to function effectively, but there must be some restrictive measures, because R&D generally has no knowledge of purchasing's problems and, engrossed as it is in its own work, has no desire to become involved.

Purchasing, on the other hand, has some strong feelings of its own with regard to what sources are to be used. Source determination is one of its most jealously guarded prerogatives, and a free rein in selecting sources tends to transfer the right of selection from purchasing to R&D. Purchasing's reasoning in guarding its right is founded in the knowledge that the corporation cannot buy from everyone, so the department spends a great deal of time developing the best sources to supply the needs of the corporation. But inherent in source development is the problem of source relations; indiscriminate use of sources poses some serious moral and ethical problems.

The most argued question today is the policy stance that purchasing should take regarding a source that has been of great help to R&D in the development stage. This problem becomes very real when the source that did the bulk of the development work is not necessarily the source that purchasing wishes to use, perhaps because the source has poor geographic location of its manufacturing units or its pricing procedures are out of line. It is not sound policy to deprive purchasing of its right to select on a competitive basis and force it to accept an unsuitable source. The problem becomes even more complex when it is the development source that came up with the recommended concept and package—more original and dissimilar than anything else submitted. In this case purchasing has a difficult job at best.

The question of source selection may not sound serious, but the effects can be most severe for the corporation. It is not unusual for a source to have faithfully produced for R&D during the development stage and then not even get an opportunity to bid on the final job—much less run it. This touchy problem causes more tension between purchasing and R&D than any other area, and a corporation's climate of cooperation can rapidly deteriorate when this contest for power starts and is allowed to continue. The innocent victims of this struggle are the salesmen for the packaging sources who are caught in the middle. They must, at all costs, keep the purchasing people on their team and yet cannot afford to offend the development group. The relationship between purchasing and packaging R&D can become so bad that effective communication between the groups is virtually impossible.

Since packaging R&D usually cannot win its battle to control the source selection techniques of purchasing, it is imperative that the head of R&D realize this early in the game. He and purchasing must work out an operational procedures formula. Basically, although few care to admit it, there is a built-in power struggle if the two groups operate under separate leadership—and R&D comes off second best because it just does not have the strategic position or authority to win the contest. Its only hope of achieving a semblance of dominance or control is if purchasing is headed by an unsure and noncombative leader. Even in this case, R&D can never really gain any measure of control over purchasing prerogatives as long as purchasing retains the power to issue the purchase order.

R&D's failure to recognize the relative strength of the positions can be almost fatal, because purchasing can, if it wishes, quickly strangle the R&D operation if no harmonious middle ground is worked out. The technique is simple and subtle. Each of the buyers quietly passes the word to suppliers: No associations are to be made with the R&D group without express permission from the buyer. With this gambit, effective R&D contact with development sources is eliminated or severely reduced. If R&D is perceptive, it should get the message that the rope is slowly tightening around its departmental neck and a mediation conference is in order. If it persists in the power struggle, the extinction process continues; this takes a little longer, but the result is just as inevitable.

The only recourse left to R&D is to use sources not presently doing business with the corporation, but these sources dry up rapidly when they discover that they are expending sample dollars and research and salesmen's time on an area that is barren of potential recoupment. Obviously, purchasing continues to buy from the regular suppliers, even to the point of gently but firmly excluding sources that R&D is using by merely not asking them for a bid. It answers any complaints by explaining that the breakdown in communications precluded its knowing of the particular source's participation in the development phase of the project. Eventually, R&D finds itself operating in a lonely wilderness and is unable to function for the corporation. The entire process is so quiet and often gentle that no one can really pinpoint what happened. The pressure that purchasing can apply, if it is forced to extreme action, is unbelievable and almost impossible to fight; sources comply with purchasing's requests because it holds the purse strings and, hence, the power.

How to avoid the whole situation. Although purchasing has strong reprisal ability, it generally abhors the prospect of such internecine conflict, much preferring to "get on with the job." The key is not to let the situation develop in the first place and, if it does, to stop it immediately. The heads of both departments must be alert and intelligent, have mutual respect, and, above all, have outstanding communication. Each should try to appreciate the problems and responsibilities of the other.

If arbitration is necessary, it tests the true caliber of these key men, because they must sit down and hammer out a set of working rules—a task that is not as hard as it sounds if both are reasonable men. In doing this, each has a chance to air his views and to understand the needs of the other. Each prepares a list of factors that he considers essential to the survival and success of his operation; then both men discuss the list on a point-by-point basis until agreement is reached. If there is an impasse on a given point, it is dropped, and they go on to the next point. As each solution is reached, it is recorded as mutually acceptable. It is probably best to have several meetings and to alternate meeting spots so that each man has some home-ground negotiating time.

Once the working arrangements have been established, the

next and most important job is to relay them to the personnel involved. Most employees react to a positive approach and to a display of harmony between leaders; therefore, a very effective technique is a joint meeting of both groups, preferably on neutral ground. This must be a command performance—no excuses for nonattendance. In conducting the meeting, the two leaders should sit together at the head of the table or on the platform. One leader should make an opening speech outlining the problems that have occurred, frankly stating the desire of both leaders to solve these problems immediately, and strongly reiterating their determination to have harmony.

Then the other leader reviews the agreement, point by point, and asks for questions from the floor. The basic need in this meeting is frankness in facing the conflict areas; this is not the time for evasive tactics or elecutionary footwork. The involved personnel will accept the program only if they realize that the leaders have worked out a sound solution that preserves the dignity and prestige of all. The keynote of the meeting must be absolute solidarity and complete agreement at the top.

The matter does not rest here, however. Any change takes time, and rules will be broken. There must be constant counseling with the personnel and insistence that violations of the agreement be brought to the leader involved. At this early stage, employees should not try to straighten out their difficulties by themselves; discipline should be exerted by the leader. The complaint should be relayed from chief to chief, and the person involved dealt with. Once a transgression has been handled in one department, it is essential that the complaining department be apprised of the action taken so that it can be reviewed with the aggrieved party. This approach gives evidence of interdepartmental concern over operating harmony.

The department heads do not have to stay directly involved in this wrangling forever, but the transition from leader action to local handling should be done gradually as an exhibition of the leaders' confidence in the ability of the men under them to follow the prescribed philosophy. Once operations begin to run smoothly, an occasional compliment to both sides about the change in climate does much to instill a pride of accomplishment.

Other packaging relationships. The emphasis has been on the relationship between purchasing and R&D, and rightly so, since R&D cannot function effectively without a good working agreement with purchasing. R&D's relationships with other departments, however, are slightly different in nature, because there is not the closeness or dependency that exists with purchasing. But there is a similarity in the fact that most of R&D's problems with engineering, marketing, design, and product development stem from communication failures, particularly disputes with engineering over mechanical equipment.

Although R&D is responsible for machinery evaluation, the bulk of mechanical knowledge lies with engineering, and its help and cooperation are necessary. Engineering has a large stake in the packaging development process and can be a valuable ally in working out problems; conversely, failure to develop a rapport results in more corporate infighting, with energies not being fully concentrated on the task at hand.

It may seem that a desolate picture has been painted regarding problems, particularly communication problems, but desolate or not, it is realistic. People are people, and to be a success in packaging management one must be able to understand them and to handle the people-oriented situations that arise.

There are those who will not agree that there are, or can be, major problems between packaging R&D and purchasing and who will resent the view that purchasing has as much power as described. But under aggressive, knowledgeable leadership, the purchasing group is a giant among pygmies. Normally, the power is latent, but it is there for use if required. The major reason for discussing the relationship is, hopefully, to forestall any future confrontations.

Chapter III

Organization of Packaging Research and Development

THIS CHAPTER details various fundamental approaches to the organizational structure of the packaging research and development department and locates its position within the overall corporation.

The most common, and apparently workable, approach to corporate location is to place packaging R&D within the confines of a laboratory setup, if the company has one. Since a laboratory is usually organized to do research and develop new products, it is normal to consider packaging R&D as a part of the overall research effort. Companies in the food industry often do this, but not necessarily general merchandising establishments or users of industrial packaging. The approach has a certain amount of organizational neatness, but often times it is not the best solution and there is a growing school of thought that takes issue with it.

Corporate Organizational Structure

In any management situation, control and efficiency of operation, shortening of lines of communication, and consolidation of functions are matters for prime consideration. The more separate agencies that become involved in any corporate function, particularly packaging, the more difficult become control and communication. In the field of corporate packaging management, where the bulk of current organizational structures still provides for independent departments not under central control, numbers lead to problems. There is a marked trend in progressive companies to recognize the complexities of packaging management and take definite steps to simplify the organizational setup for better control and, hopefully, better results.

Several general merchandising houses have, after struggling with the problems of cooperation and function among packaging's big three departments, reduced them to two—design and purchasing/development. The merger of purchasing with packaging R&D is logical and practical because the ties between the two are so close. The mere fact that R&D depends on purchasing to a great extent for technical help, source development assistance, and, often, for exposure to new developments almost makes a combination department a necessity. The plus factors in merging the two are many. Experience has shown that it results in:

- Increased and, generally, better quality output by R&D.
- An increase in the technical knowledge levels of employees because of better exposure to the technical staffs of sources and easier access to the vast store of knowledge available in the average purchasing group.
- Complete elimination of interdepartmental tensions and problems.
- The development of more technically professional people who have an increased awareness of the importance of their function in the corporate whole.
- Reduction of communication and control problems.

Many companies that are heavily product development and

laboratory-oriented, such as those in the food industry, have never given serious consideration to the department-merging concept. It is not difficult to see why. There are some fundamental psychological barriers. People in the business world tend to defend their kingdoms regardless of the corporate well-being; therefore, such a departmental consolidation means a major corporate structural change and it precipitates some monumental battles. Top management must overlook the roars of indignation and focus on what is best for the company. Generally, the management of the purchasing group is stronger and more decisive than that of R&D. The type of activity that purchasing is involved in requires strong, decision-oriented people, whereas R&D's personnel more closely resemble the general concept of the researcher, although in fairness it must be said that packaging researchers do not fall into the category of the average research scientist but are a combination of two personalities—science-oriented and practical-minded. By their contact with the marketing, purchasing, engineering, legal, and other departments, they stay within the frenetic atmosphere of the corporate world. Nevertheless, they must operate more slowly and carefully than other areas because if anything is wrong with the package they developed they suffer the full wrath of top management. Knowing this, the average package development man has a built-in reservoir of suspicion until everything has been thoroughly tested. His is a world of doubting pragmatism.

Departmental Organizational Structure

There are some basic guidelines to keep in mind when approaching an organizational revision:

1. Review the organization of the entire corporation carefully, particularly the product or merchandise groups.
2. Evaluate the relative importance of the various subdivisions of packaging, such as shipping containers and testing, the basic types of retail packaging, available packaging equipment and its effect on the development area, and mail-order packaging (if any).

3. Establish exactly where packaging R&D is to fit into the corporate setup.
4. Determine the amount of service to be given to packaging R&D by other departments and, consequently, the amount of control these other departments should have over certain aspects of packaging decisions.
5. Carefully study the purchasing setup and check for problem areas.
6. Prepare a solid working relationship with purchasing, packaging, design, and engineering.
7. Review the design setup. Are its men presently involved in package development? If so, prepare a department structure that will get them out of the development area.
8. Review all the men involved in packaging R&D carefully and analyze their talents. Use every means possible to ascertain weaknesses, strengths, and interest areas.
9. Establish a philosophy of operations.
10. Develop a workable plan for operations.

These are but ten checkpoints to consider. There may be more, depending on the individual circumstances, but if a manager uses these ten, plus improvisations, his organizational setup will be off to a solid start.

Although there are numerous ways to organize an R&D department, there are some common factors among the variables. For example, the titles bestowed upon the functionaries of packaging R&D vary according to the preferences of the corporation, but the basic structure varies little, regardless of the leader's reporting channel. Depending upon the size of the department, the basic structure looks something like the following:

- Manager (one).
- Project or team leaders (usually not less than two).
- Packaging technologists (four or more).
- Specification coordinator (one).
- Secretaries (two or more—usually report to team leaders).

The project team's orientation depends on the philosophy of the corporation and the manager. Some companies will lean toward the packaging-industry orientation, while others will have a merchandise or product-unit disposition. Both of these approaches are fairly common in the food industry, but the product or merchandising group alignment seems to be gaining favor. This makes a great deal of sense in terms of involving packaging R&D more deeply with merchandising, and it aids in shortening the lines of communication. But there are some who believe, and with justification, that a packaging-industry orientation has more to offer. The reasoning behind this is that, with diligence, the technologist becomes expert in his chosen field, such as cans, folding cartons, containers, or plastics. The major disadvantage of the concept is that it tends to inhibit the knowledge range of the individual and to reduce flexibility within the department.

Advocates of the industry orientation point out that the merchandising group approach tends to produce jacks-of-all-trades instead of experts. Although this is true, it is not a fatal flaw. The lack of outstanding expertise is far offset by the reduction of communication problems with marketing and by a much greater awareness of what competitive packaging is doing in the marketplace. The more the technologist knows about the plans, hopes, marketing problems, manufacturing capabilities, and machinery developments of the entire industry, the more expert he becomes in relevant areas of packaging and the less he becomes involved in nonpertinent packaging materials and techniques. In a sense, he becomes an expert in the fields in which his industry operates. The organizational chart for the merchandise- or product-oriented group would resemble Exhibit 1.* The interest areas are only intended to be representative and vary according to the product lines of the corporation.

*The reason "equipment" is listed in several places is that each product demands its own packaging equipment to form, fill, and seal. Rarely can a major packaging innovation be exploited without development of equipment to handle the new package. Take, for example, the case of a product that needs a paper and film combination pouch. Pouching requires the development of hitherto unavailable machines to fill and form the pouch at high speeds and other machines to set up the folding cartons and load the pouches.

Exhibit 1

Merchandising Approach (Food Industry)

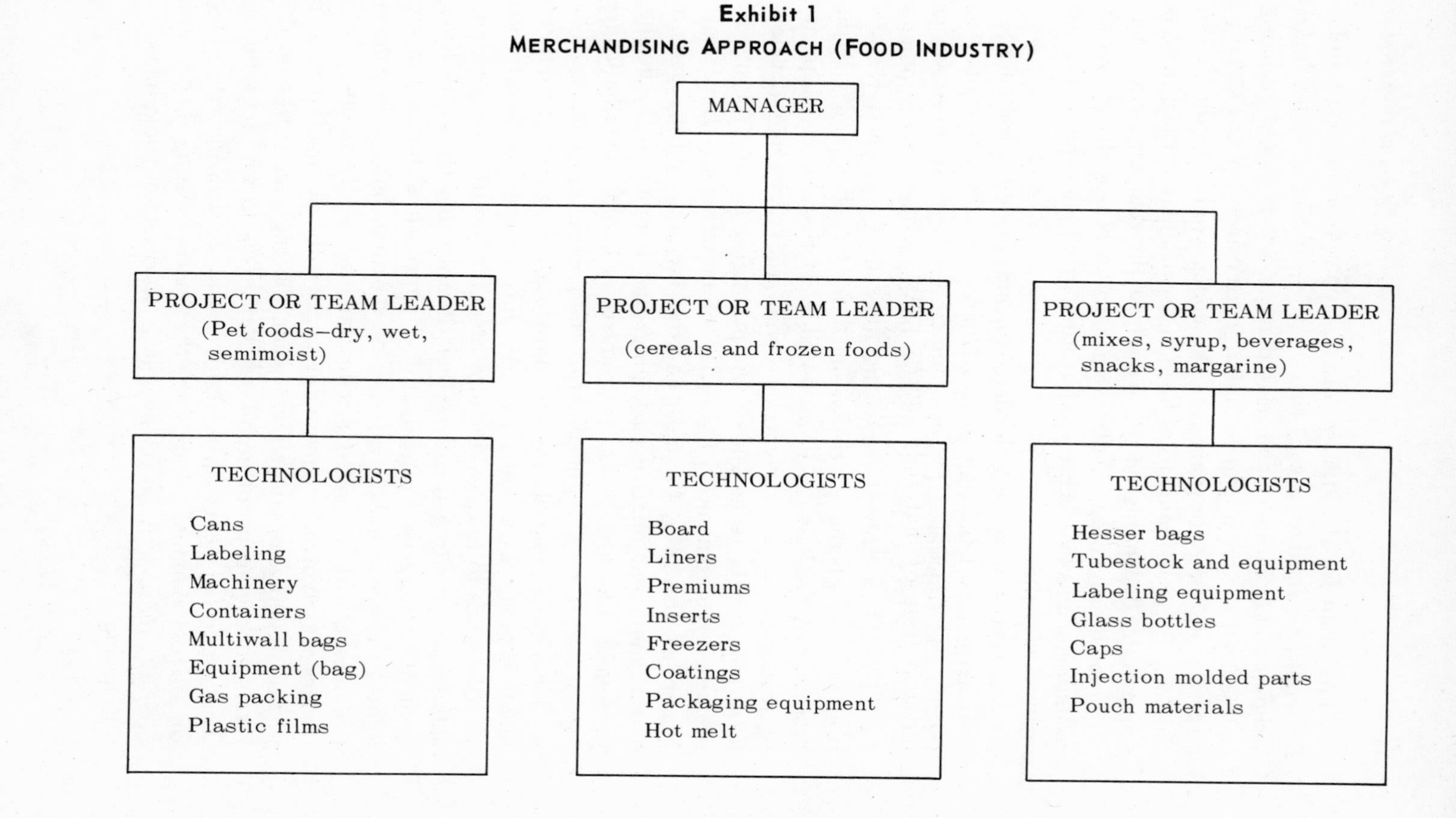

The setup for the industry-orientation approach (see Exhibit 2) would be similiar except that there would be a technologist assigned to key industry and related areas. There would probably be less need for team leaders and a greater need for one or more direct assistants to the manager. These men would coordinate and control project assignments. The industry-oriented arrangement is expandable, depending on the growth of the company, but it has a serious disadvantage in that it greatly increases the communication problems between the laboratory group and the marketing or product people.

For success in solving the departmental organization question in general merchandising houses, there must be a blending of talents and functions. The training of the men involved varies and often serves to set the pattern of department development. In this field, industrial packaging must be given special attention because of the particular need to get the goods to the customer in the best condition possible. If there is a multiplicity of merchandise departments involved, the packaging development men prefer to handle all the packaging needs, exclusive of container requirements, of a group of departments. Experience has shown this to be the most effective method, and it results in better, more imaginative packaging, and happier, more flexible personnel. The men prefer the versatility of different challenges to working in only one medium of the packaging spectrum. The tendency to pigeonhole the men into hard- and soft-line departments is being discarded slowly, because to restrict them to one activity tends to relegate them to certain types of packaging and acts as an inhibiting factor to their interest and technical development. A suggested organizational approach is outlined in Exhibit 3. There is quite a mixture of function, but the alignment provides a wide range of intellectual interest. If shipping containers are involved, many of the areas will fall into the domain of the shipping-container team while the engineer works on the retail packaging problems of smaller items in the department. This suggested approach is highly workable, has the advantage of direct communication with the merchandising units, and operates quite well on its own if it learns to accommodate to purchasing.

Exhibit 2

INDUSTRY APPROACH (FOOD INDUSTRY)

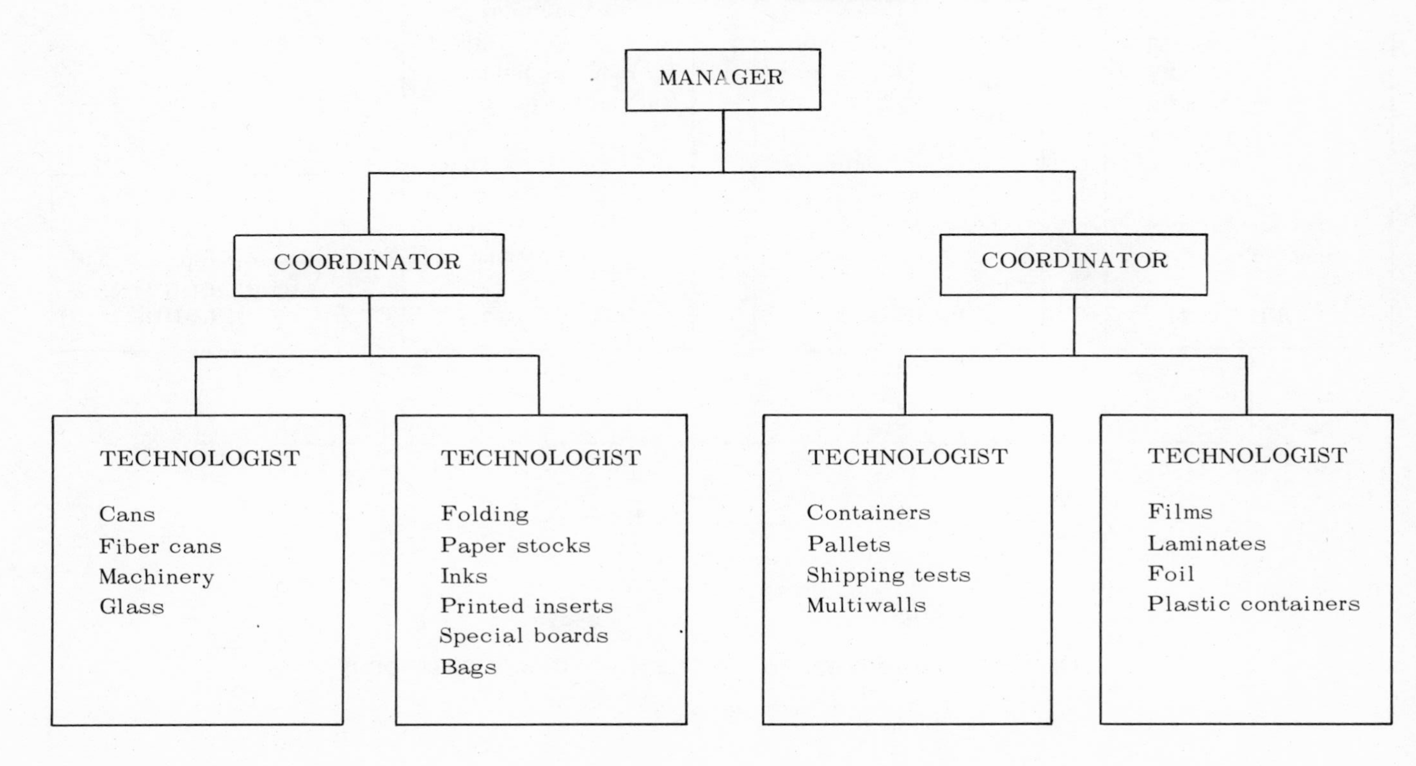

Exhibit 3

SUGGESTED APPROACH (GENERAL MERCHANDISING INDUSTRY)

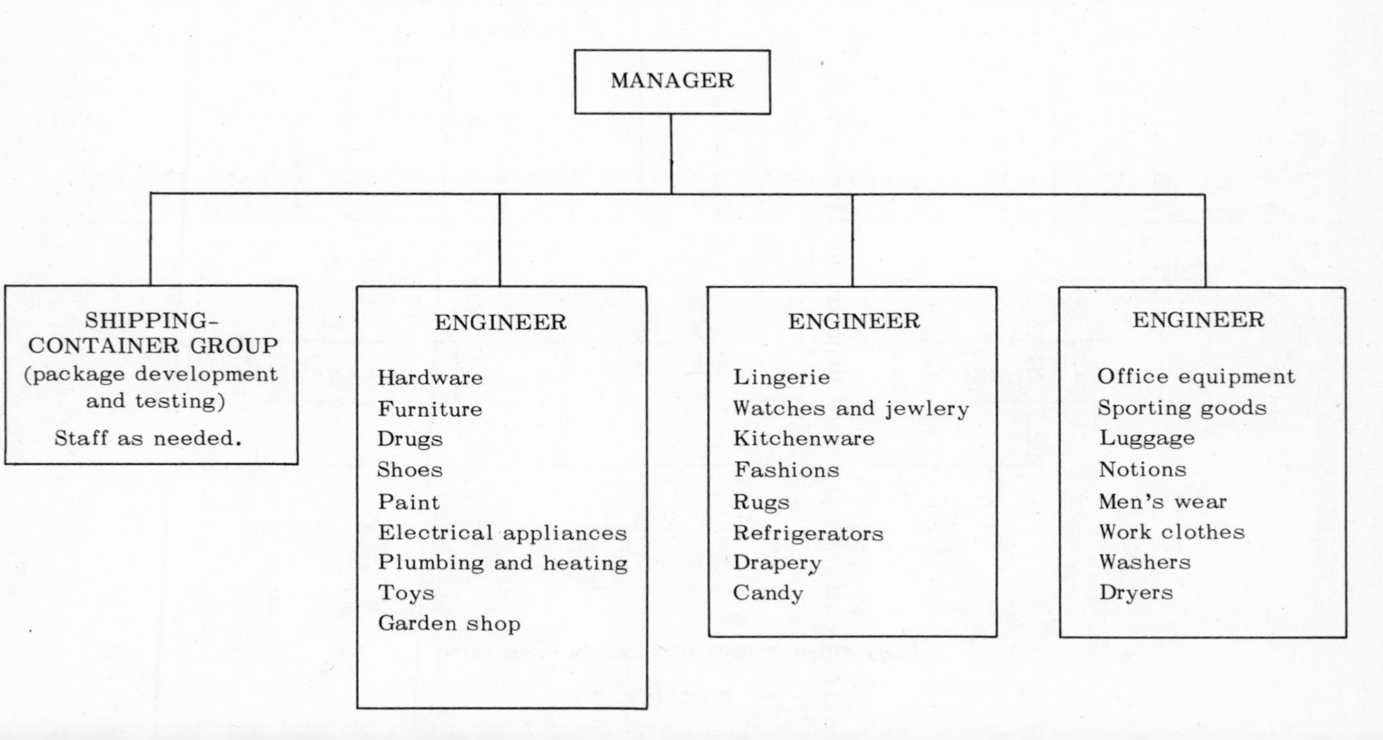

But it works even better if it is part of the purchasing team. The purchasing and development men, when combined into one department, spend less time jousting for position and more energy on accomplishing tangible results.

Reporting channels. The direct reporting channel of the R&D group varies slightly, depending on where in the corporate structure it is situated. If it is aligned with the laboratory and reports to the laboratory hierachy, the direct channel will probably be to a director of a research area. On the other hand, if packaging development and its equipment are large items of expense, there will probably be a separate director to whom both groups report. If it is a part of the purchasing setup, R&D report to the vice president of purchasing as a distinct responsibility. Since package purchasing is the dominant function of the purchasing effort, it is possible that R&D could have a straight-line relationship to the director of packaging purchases. This is a bit neater and tends to build a strong team. Exhibit 4 shows how this chart might be set up. An alternative would be to have the manager of R&D report directly to the vice president as a third director.

Personnel requirements. The task of selecting a department

Exhibit 4

PACKAGING R&D REPORTING TO THE VICE PRESIDENT OF PURCHASING

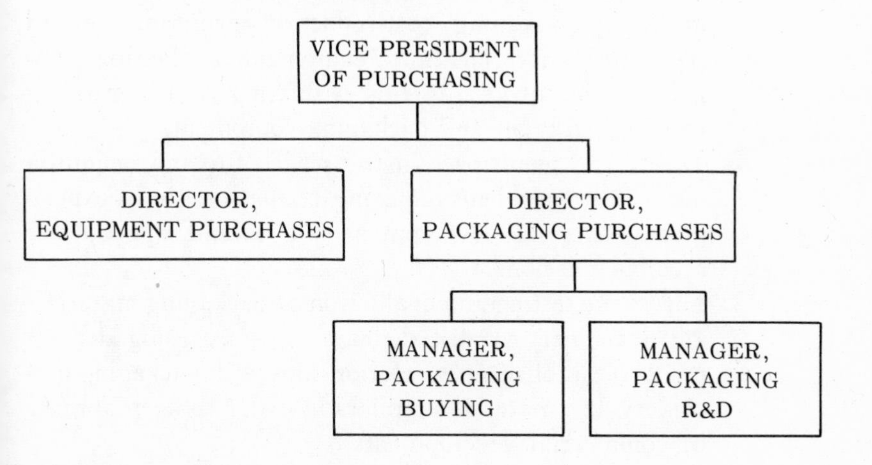

manager and technical personnel is most difficult because of the shortage of persons having the background and experience required by the jobs. But the first step is the job description; if it is properly and carefully made up, a great deal of time can be saved in personnel searching. It should clearly outline the accountability objective (basic function), nature and scope (major responsibilities), and the principal accountabilities (general, summarized responsibilities). Following is a suggested job description for the manager of packaging research and development. It is only a model and should be used as such.

JOB DESCRIPTION FOR MANAGER OF PACKAGING R&D

Accountability Objective

Plan, direct, and control packaging development activities in both mechanical and materials fields and coordinate the interdepartmental packaging efforts of the company.

Nature and Scope

1. Develop and recommend to the lab director and director of packaging purchases long-range packaging plans, including packaging cost-reduction programs, broad packaging materials and equipment evaluation programs, and programs for evaluating and improving existing packages and packaging equipment.
2. Direct and participate in the search for and definition of technical problems requiring evaluation, study, experimentation, and development and submit appropriate recommendations.
3. Direct the testing and evaluation of packaging materials and machinery. Work through the purchasing department to establish close relationships with packaging machinery and material suppliers in order to keep abreast of commercial developments.

4. Work with the product development research group and conduct necessary additional tests to determine newly developed products' packaging requirements. Direct complete development of the package, except for artwork and copy, as follows:
    a. Direct the determination of required packaging materials specifications and, in consultation with the marketing department, determine package size, form, and materials to be used.
    b. Evaluate with the production and engineering departments and with the manager of engineering development the relative advantages of using existing packaging machinery against purchasing (or developing) new packaging machinery. Make recommendations accordingly to the director of packaging purchases.
    c. Direct the development and testing of experimental packages.
    d. If the package is to run on existing packaging machinery, work with the appropriate production manager on directing the production-line testing.
5. Make recommendations as to the selection, discharge, and evaluation of packaging development personnel.
6. Direct the development of new packaging machinery and processes and the improvement of existing machinery and work with:
    a. Production, engineering, and purchasing and with the manager of engineering development on the testing and evaluation of available commercial machinery. Inform the director of packaging purchases of the need for equipment or of the development of new or additional equipment within the company.
    b. Production and engineering on adapting improvements to existing machinery or to that developed within the packaging development organization.

7. Work with the engineering and production departments to analyze packaging material or equipment that is not producing packages of specified quality or at specified production rates and provide assistance in making improvements.
8. Keep alert to packaging activities where improvements or cost reductions appear feasible and, within budget limitations, direct exploratory work. If exploratory results indicate that improvement is practical, recommend to the director of packaging purchases that funds be appropriated for further development work.
9. Direct the preparation and development of the package and the packaging materials specifications and standards.
10. Provide necessary liaison and information flow among all departments that have an interest in packaging in order to supply them with current information on industrywide and company plans and developments and to insure that all packaging ideas originating in the company are considered, evaluated, and, if appropriate, recommended to the director of packaging purchases. Also, serve on the corporate packaging committee (if there is one).
11. Participate in start-up operations of commercial processes or modifications originally developed by the packaging development organization to advise and assist the production and engineering departments.
12. Make recommendations for the growth of a packaging development organization of sufficient strength and ability to deliver all normal services in the areas of assigned responsibility.
13. Prepare and recommend budgets for packaging development activities and control expenses in accordance with approved budgets.

Principal Accountabilities

1. Direct the long-range packaging development plans for packaging equipment and materials.

2. Direct problem-solving efforts in technical areas of packaging and machinery.
3. Direct development and testing of packaging for new products.
4. Direct materials and equipment evaluation studies.
5. Direct preparation of specifications on materials for the purchasing department.
6. Direct coordination between the development group and the merchandising, engineering, production, purchasing, legal, and design groups.
7. Direct the development of new machinery and processes for the plants.
8. Direct the development of new test methods and refinement of current techniques.
9. Direct all shipping tests where packaging is concerned.
10. Prepare the personnel plan for the department.

The manager reports to the director of (whatever title is used in the organization) and supervises project or team leaders, technologists, and clerical help.

Desirable Characteristics

Once the job description has been completed, the next task is to fill the job with the right man. Some definite requirements should be:

1. Ability to get along with people. The amount of personal contact work involved in this job demands a man who is people-oriented. One who is not is virtually doomed to failure.
2. Ability to communicate clearly and effectively. A good "floor presence" is a great help in presenting projects and ideas. If the applicant fulfills all the other requirements but is not too adept at public speaking, he should get training.

3. Ability to guide and motivate people.
4. A technical as well as liberal arts background. The technical area could be chemistry, physics, engineering, or a related field.
5. A work background of at least ten years in the packaging field, preferably in research and development.
6. Ability to plan and to get the job done. Often, the research-oriented man has difficulty arriving at a final decision. But the man who is to head packaging R&D must be a decision-maker.

Undoubtedly, there are many other desirable characteristics, but as a basic guide these six will do. They are must requirements on which there can be no deviation; however, they are also rather demanding. Where will such a manager be found? If the corporation presently has a functioning packaging R&D group, care should be given to the development of a home-grown manager. Promotion from within is usually a fundamental of corporate life, and the need to go outside for a new manager suggests a massive failure in the personnel development program of the corporation. But, assuming that there were no in-plant candidates, where would be the logical search locations?

1. R&D managers in allied industries.
2. Outstanding nonmanager personnel in allied industries.
3. Research personnel currently functioning in the research activity of packaging suppliers.
4. Highly skilled packaging buyers.

JOB DESCRIPTION FOR PACKAGING TECHNOLOGIST

The task of selecting packaging technologists is somewhat easier. After carefully completing the job description, much of the work is done. A typical job description is as follows:

Accountability Objective

Conduct research and development studies for evaluating packaging materials and methods and originate and search for new packaging concepts.

Nature and Scope

The technologist is primarily concerned with developing tests and executing studies to determine the feasibility of new, improved packaging concepts and materials and for making recommendations to management through specifications, correspondence, and meetings. The studies are finalized by his personally conducting physical tests of materials and by evaluating the results of consultations with other R&D groups, marketing, engineering, and purchasing, plus outside sources. The technologist is to participate in plant start-ups when new packaging concepts or materials have been adopted. He must be aware of, and keep alert to, the innovations and changes of the packaging industry, and he must develop and maintain close personal relationships with key product development and marketing department heads and with supplier companies. Self-motivation, imagination, technical packaging knowledge, and personal salesmanship are required.

The technologist, who reports to the manager of packaging development, is to act within limitations established by the manager and confer with him on general matters of project performance and procedure. He is to work closely with the packaging buyers in the utilization of sources for packaging development projects.

Principal Accountabilities

1. Originate or search for new packaging concepts that could be adopted by the company.
2. Ascertain the feasibility of various approaches and solutions to packaging problems.

3. Conduct tests on approved projects to develop new and improved packages to meet marketing's needs.
4. Advise the purchasing, marketing, engineering, and production departments, as well as other research groups, to utilize optimum technical knowledge in completing packaging production plans for various products.
5. Instruct plant personnel at start-ups of new operations on the proper use of new equipment and materials in order to minimize materials and manpower waste.
6. Develop new packaging materials test methods for laboratory and quality control use.
7. Keep aware of modern consumer packages, materials, and test methods through literature, personal contact, supplier visits, and observations to insure company of up-to-date concept, quality, and economical packaging.

The "hunting grounds" where corporate recruiters can find such technologists are somewhat wider than for the manager because the experience level can be reduced in view of the amount of in-plant training that will be required:

1. Packaging school or engineering graduates.
2. Packaging development men at packaging sources.
3. Experienced assistant buyers of packaging.
4. Packaging engineers in almost any package-oriented corporation.
5. Packaging salesmen with a technical bent.

The efficacy of the packaging research and development function greatly affects the marketing efforts of the corporation. Therefore, it is worth top management's time to make sure that it is well organized, well staffed, and operating at peak performance.

Chapter IV

Packaging Design

PACKAGING DESIGN is the second of the big three groups in the corporation's packaging management complex. It is vitally important to the success of the corporate packaging effort and is an operation that can cause considerable trouble unless properly established, supported, and controlled. The various activities will be explored in detail after the group has been placed in the corporate hierarchy.

LOCATION IN THE CORPORATION

Although the "where" and "how" of locating the packaging design group often presage its success or failure, there is a great deal of diverse thought on its reporting status. Experience has shown that to be fully functional in a packaging-oriented corporation, and to give weight to its decisions, the group must report to a senior executive.

Ideally, the design director should have the last word on what is good design, but in practice this is seldom the case, unless he has strong senior-executive support and a clearly and distinctly defined area of authority. The optimum arrangement would have him report directly to the president, but, unfortunately, it is usually his fate to report to the senior merchandising or marketing executive, who also controls the marketing personnel. When this happens, the system of checks and balances so necessary to the smooth functioning of the overall effort is removed, and instead of having one design director, the corporation ends up with every marketing man a design expert. The one real expert, the director, becomes a diplomat without portfolio. The effect of all these pseudo art directors creates havoc on schedules, on the exploration of ideas, on the creation of workable and reproducible designs, and fathers a Pandora's box of troubles for the purchasing department. When eager amateurs try to appraise design and printing, confusion becomes the order of the day; yet the list of corporations whose design concepts are at the mercy of eager but uninformed neophytes is legion. The only salvation is either a tough design director or an extremely knowledgeable and hard-nosed purchasing department.

On the other hand, if the company's top executives are enlightened enough to realize the value of a good design director and the saving grace of a system of checks and balances in the corporation, the director will have presidential backing. The mere fact that he becomes, in effect, part of the president's staff serves notice to marketing that his decisions must be reckoned with as expressions of opinion from the top. But even if he and his staff are properly situated within the corporate structure, the burdens that fall on the director are immense.

The first thing he must do is recruit a staff capable of doing the work—both in quantity and quality. Just as for R&D, the job specifications must be explicit in outlining the duties of each man. His second and most difficult task is to build his job and his stature within the corporation; he cannot be made successful by executive mandate. He must earn, as do all managers, the right to authority in the corporate structure. In many respects, his spheres of activity are more difficult than those of packaging and purchasing. These two departments deal with concrete, tangible, physical objects and situations with clearly visible parameters; his department deals, to a great extent, with ideas and opinions. Designs or colors or copy are likely to be areas where everyone has a definite opinion. Add to this the fact that each marketing man or merchandiser has or should have a great store of specific knowledge on his product and the makings of bitter controversy are present. The design director treads a dangerous path between what he wants or believes necessary in design and what the marketing people want.

Marketing has a definite responsibility for alleviating the potential conflict by preparing a well-developed concept of the project, which will lay the groundwork for the design. The lack of adequate concept analysis is probably one of the greatest weaknesses of marketing personnel; however, it is closely followed by their apparent inability to finalize a decision within the specified time requirements and the inherent belief of all marketing men or merchandisers that they know more about what is best in design than anyone else. Thus the picture of the average design dilemma is clearly outlined.

Faced with these ready-made problems, the design director must embark on a personnel-selling program, at the same time delicately concealing the mailed fist of final authority. He must establish himself in the eyes of all the merchandising people as the ultimate authority on what is truly good design and yet is commensurate with the corporate objectives. He must be knowledgeable, persuasive, firm but gentle, tolerant of the opinion of others, able to compromise and bargain where needed but able to hold his ground when challenged—in short, he must be a man of unusual breadth of experience, talent, and character.

Packaging Design Functions

Packaging design is responsible for a wide variety of activities including:

1. Control of the corporate design policies for all products.
2. Development of design concepts from sketches through comprehensives to final artwork.
3. Coordination of all related activities concerned with a complete packaging development project including
   a. Maintenance of schedules with the merchandisers involved.
   b. Coordination of schedules with packaging R&D, purchasing, engineering, and production.
   c. Clearance of all copy with the legal department.
   d. Preparation of all projects for necessary approval.
4. Development of corporate packaging design objectives.
5. Development of appropriate design sources to meet packaging objectives.
6. Checking of technical packaging requirements (conformance to corporate color standards, tolerances on printing runs, and approval of final color standards).
7. Preparation and administration of the packaging design budget for the corporation.
8. Development of the concepts and preparation of the artwork for all packaging changes required by the merchandising problems of the marketing groups.

Design control. One of the primary areas of activity is the control of the corporate design policies for all product lines. Depending upon the corporation's philosophy, this activity may or may not be the cornerstone of the packaging design operation, but the increasing emphasis on corporate image, product identification, and the interrelationship between products favors its being so.

Package design is involved in the development of corporate logos, designs for packages, product-line color programs, product-promotion approaches, and, on occasion, even corporate letterheads, so the development of a corporate design concept is complicated and it requires not only strong initial planning but solid execution and careful follow-up on future package-design developments. There are no real rules to the game, but certain fundamental characteristics usually evolve, such as:

- The development of either an easily identifiable corporate logo or a variety of subcorporate or product-line identifying symbols, or both.
- The development and use of distinctive colors or groups of colors for easy identification of given product lines.
- Skillful blending of these identifiers in advertising.
- Careful coordination of the key design elements in location on the package, color, and type matter to further customer identification of the corporate characteristics, regardless of the media used.
- Careful selection of colors to convey the desired impression upon the customer. Colors must blend with an emotional need of the customer.
- Development of design concepts aimed at maximizing use of the specified packaging materials and the physical shape of the package.

Design development—tools and techniques. This area is crucial to the corporate package development program because of its responsibility for surface graphics. Moreover, it is a necessary preamble to the work of the purchasing people because they must have camera-ready, finished artwork before they assign suppliers.

A project's design development program has definite phases. Initially, marketing and packaging R&D must come to terms on the rules, regulations, hopes, and desires to be followed. Once they have established the limiting parameters, design's work can begin.

When design receives the physical shape of the package from R&D, it first formulates a plan for the design. Much of this plan, which is really a reiteration of the image that marketing wants to project, has already been formulated in the preparation of the marketing base or concept. The design director and his staff are to translate the image into tangible items, such as characters, shapes, colors, and functional-parts arrangements; they also must have a plan for future changes in the package. It is here that the director's knowledge pays off—he must be aware of colors and their connotations, of printing techniques and limitations, and of the basic concepts of good design. When the plan and instructions are prepared, the director will probably start consultations with the design studio or studios that will actually create the required designs.

Design studios. The design studio is the major working tool of the corporate design director, but its effectiveness depends upon his ability to visualize the package concept, transmit the nuances of what is desired, and control the studio's actions. The last is probably one of the areas least under the corporation checks and balances system. The director controls the budget, selects the studios, and directs their action with practically no interference from anyone.

The studio will probably prepare the initial idea sketches, final sketches (if required), comprehensives, and the final art-work. The development of a design generally encompasses the following steps. First, the director meets with the marketing or merchandising people to develop the basic requirements for the package, covering such general items as marketing base, image projection concept, segment of the market to be attracted, color relationships within the corporation and within the line related to the new item, marketing schedules, design schedules, copy approach, and future promotion ideas. As mentioned, unless the merchandisers have carefully researched their market and placed their item in the proper segment, they will be unable to answer the required questions. There is no way of knowing how many bad packages, resulting from poor marketing planning, have been blamed on the design group. Planning, particularly of new packages and new markets, is hard work, and there is often a tendency to toss the project to design and just wait and see what develops. This type of approach is dangerous, expensive (studios get paid for all the time spent and sketches created), and, in the final analysis, rarely satisfactory.

The next step is to brief the studio or agency thoroughly on the project. Whether the merchandisers are in on this session depends on the warmth of the relationship with design—a relationship that is always tenuous at best, which is why many directors prefer to be the key contact with the agencies. One school of thought feels that purchasing should be included for consultation. The reasoning is that purchasing managers and buyers often suffer when packages are designed without proper consideration for the reproduction problems at the manufacturing stage. However, purchasing rarely is invited to consult because the design director feels that if it ventures an opinion as to number or type of colors, or on any other design question where a technical problem may be involved, the creative capablities of his group are being hampered. Designers feel they should be unfettered, unhindered to create "freely," but if they would heed some advice from purchasing, a great deal of later trouble and controversy could be avoided. Still, the most important point at this juncture is that only one set of instructions is worked out—contradictory instructions are fatal.

When the initial sketches are received from the design studio, they must be reviewed with the merchandising group. At this meeting or series of meetings, purchasing should be included and the ideas represented by the sketches should be reviewed, discussed, rejected, or accepted. The normal activity is one of compromise, cutting, choosing, adding, or deleting. Hopefully, the design desired can be jelled at this point and the proper guidance given to the design source.

The next item on the agenda is the comprehensive, which is nothing more than a mock-up of the package configuration with the main areas of color blocked in and related to location of copy areas. This is generally an important step, particularly to the merchandise groups; people not active in the technical side of packaging often have difficulty visualizing a package from sketches. The comprehensive shows the basic package appearance. The colors will probably not be correct, but at least the overall concept can be seen and judged. This stage should be marketing's last opportunity to make changes. As the comprehensives are being reviewed, the copy and the photographs or illustrative shots to be used should be studied, etched, revised if needed, and approved. It is here that purchasing people are most valuable, not as art critics, but as experts who can easily identify areas that will not reproduce as envisioned. Review of the comprehensive and the photography should be a standard operating procedure that forewarns all involved of possible hitches in final production. This is the time to reshoot a photograph—*not* after the final art has been approved, plates produced, and the first proof pulled. If the photography is wrong, no amount of press adjustment or mechanical touchup can make it right, so if marketing does not like the photo it should have design reshoot it. But purchasing must insist that there be no major alteration of the existing artwork.

Final artwork, which is also called black and white or mechanical art, is the layout of the package with all the components in their proper places. It is the standard toward which the printer works. Accompanying the final art must be transparencies or die transfers of all photographic reproductions appearing on the package, a color flap showing the areas occupied by the various colors, and color swatches that are large enough to be successfully matched. Swatches should be no less than two by three inches (three by five inches is preferable); normally, the color swatch is too small to be usable. The design director often attaches a special set of production instructions that are pertinent to any trouble spots contained in the artwork. The more complete the art, the fewer the problems.

The next phase is the color proofs, which should be an accurate reproduction of the artwork. Design directors have fostered the common misconception that the proof becomes the printing standard. This is technically not true, because proofs are made one color at a time on a proof press, and since each color is allowed to dry before another is laid on, the colors are generally more intense than production colors, which are laid on wet. Wet-trap colors have a tendency to blend a bit; therefore, the final shading will be different than that on the proof. This problem should be pointed out by the purchasing people, but it can be somewhat alleviated by the use of wet and dry color charts.

The proofs are usually reviewed by all concerned. The art director and his production assistants are present along with the members of the merchandising group involved. Purchasing generally sends both the buyer and the source involved. The technical advisors from the source are extremely valuable at this meeting since they can explain the printing problems and can give a realistic appraisal of what to expect in pictorial reproduction, color control, and run consistency.

Color standard approval. In keeping with the theory of a system of checks and balances, the design director and his staff should have the final say on the color standards when the job has been run. Color standards are limitations on printing variances—that is, they maintain color quality. Every printing job has a range of color intensity, regardless of the printer's equipment and personnel capabilities, and careful examination of any sheet shows variations in the ink lay from one edge to the other. Purchasing's job is to minimize these variations by selecting sources with the best in equipment and personnel, by establishing the maximum tolerances and the color standard or objective before the run starts, and by setting up checking procedures with the source for constant monitoring of the run. This problem of color variation becomes even more acute if the corporation is extremely quality conscious.

As a rule, the standards are selected from the light and dark extremes of the first production run on the new package. This is the easy way, but it is the least effective since the source bears the burden if the ranges are unsatisfactory—a situation that often leads to controversy between the source, the design director, and purchasing. It is unfair and destructive to good relationships to put this burden on the source, since the decision on whether a color is good or bad is basically a matter of individual judgment. Theoretically, the design director would reject the run by not approving the color standards, but since there are often considerable sums of money involved and schedules to meet, purchasing would invariably contest this decision.

Some enlightened purchasing departments have developed a system of pulling standards artifically. The pressman gets up to a standard that is approved at press side and then sets the light and dark tolerances by adjusting the ink fountains on the press. This is time consuming, somewhat tricky, and expensive because it adds to the makeready costs. However, it is realistic; when the tolerances have been established, the press crews have a set of color ranges available for constant reference. The system also provides a safeguard for the source—it knows what is definitely required—and provides purchasing with specific grounds for acceptance or rejection. The key to success in this approach is to have a representative from purchasing, one from design, and one from merchandising at press side to set the tolerances. If these people cannot be there, then the decision must be made by the purchasing representative, who should always be present at the first run on a new package. On repeat runs, representatives should not be needed since the source will have the standards.

Coordinating functions. Design usually handles the liaison work needed to get a new package developed and to effect changes on current packages. It does this through its coordinators, who are generally assigned to specific product or merchandise areas. This is essentially a scheduling and contact activity, and it is most vital to smooth-flowing packaging development.

The coordinator becomes involved with the package at the outset. He is responsible, and should be, for insuring that all the facets of the complicated schedule are established, maintained, and updated as needed and that all agencies that are assigned tasks complete them on time. This is difficult because he serves as a staff man and has no line of authority over any of the people with whom he works. To do his job properly, he must check, consolidate, and activate a variety of interest spans. The areas vary by the type of corporation, but the following representative checklist is indicative of the wide scope of the coordinator's responsibility.

Merchandising. Contact the merchandise group involved to determine the schedule for:

- Consumer reaction development (CRD) testing, if any is to be done.

- Test-market ready date, if a test-market approach is to be used.
- Advertising on television, on radio, and in newspapers.
- Shelf-introduction date for the new item. This date is generally related to the advertising-break dates. However, advertising must be bought and locked-in well in advance of the shelf or production date, so the coordinator must also know the last date that advertising can be withdrawn if necessary.

Production and engineering. Both of these departments should be checked for start-up dates and any necessary equipment lead times to counter the possibility of a communication gap between the merchandise and production groups.

Purchasing. Needed here is the date when purchasing must have the final, correct, camera-ready artwork to insure that supplies are on site and on schedule.

Packaging research and development. The date must be established for the issuance of final packaging specifications. It must take into account any needed testing, such as that for quality preservation, shelf life, machinery-operating problems, and a potential host of others. This date must coincide with, or preferably be prior to, that set for the delivery of final artwork to purchasing.

Packaging design. Within his own department the coordinator is responsible for developing dates to cover the following critical items:

- Arrival of the physical shape of the package from R&D.
- Delivery of initial sketches.
- Submission of revised sketches if necessary.
- Final approval of sketches.
- Arrival of the comprehensive and a time allowance for meetings, revisions, and approvals.
- Receipt of advertising copy and any mandatory legal copy, such as formulas, ingredient listings, warnings, and the like, to be used on the package.
- Final approval of copy, allowing for rewrite and legal department check and approval.

- Shipment of CRD test materials if needed. This is worked out between the CRD test section and purchasing, with the coordinator acting as moderator.
- Completion of all necessary approvals.
- Delivery of final artwork, allowing some time for revisions and changes.

Obviously, the coordinating task is complicated. Not only must the schedules be established, but the coordinator must constantly check to keep them on time. There are so many areas involved that a breakdown in function, a delay in making a decision, or a failure to meet the schedule for any reason can seriously handicap a corporation. In some industries, particularly food and seasonal, a missed introduction can mean being out of the market for perhaps a year, and a lost year in the marketplace cannot be recovered because competition catches up and development dollars may well have been wasted.

Chapter V

Organization of Packaging Design

THE PROBLEMS of managing this department break down into several distinct areas, such as communication within the corporation, development and use of available design sources, and scheduling to complete the job on time. Job titles have very little to do with the structure of the packaging design organization—it is the function that counts. A typical department setup would include a manager, an art director or supervisor of graphics, packaging coordinators (the number varies according to the number of merchandising groups needing service), and, if needed, a copy chief.

The group's organization generally follows the same pattern in both food and general merchandising fields. A typical departmental organization chart would look like Exhibit 5. As the department grows, more assistants are added. The structure is

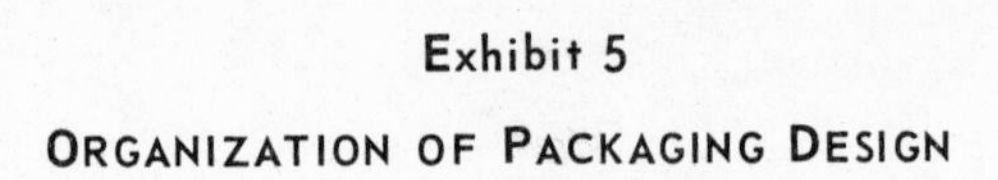

Exhibit 5

ORGANIZATION OF PACKAGING DESIGN

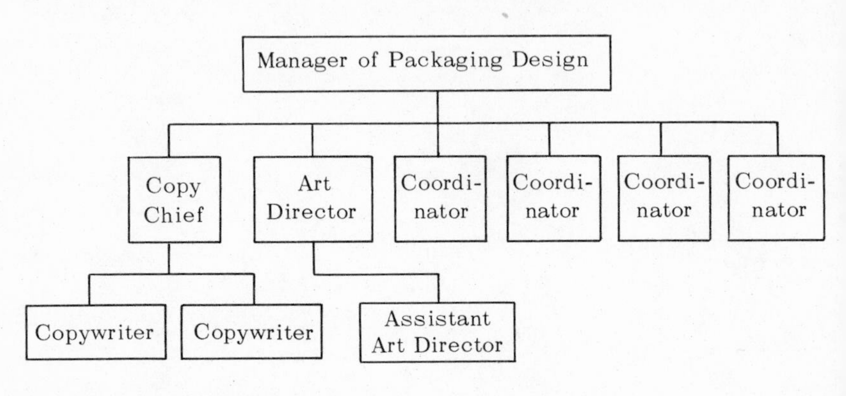

simple, with straight line reporting, and it should be kept as clean as possible.

JOB SPECIFICATIONS

Drawing up job specifications for packaging design personnel is difficult, since many of the functions defy abbreviated prose definitions, but it is vital that the jobs be accurately described and circumscribed. Too often, particularly in the dealings of the packaging coordinators, there is the danger of overlap into packaging R&D. Care must be taken to preserve the integrity of each of the major operating groups.

The major problem in creating job specifications seems to be in determining the approach and scope of the positions. The following specifications for the manager of packaging design, the supervisor of packaging graphics or art director, the packaging coordinator, and the copy chief are suggested approaches to defining the jobs.

JOB DESCRIPTION FOR MANAGER OF PACKAGING DESIGN

Accountability Objective

Manage the timely development of packages that will meet the market and corporate objectives.

Nature and Scope

The manager's concern is the development of a companywide packaging program that provides total packaging design service to the merchandising groups. To do this, he works closely with these groups to guide and assist in establishing packaging objectives that are consistent with the needs of the corporation and that serve to tie together the identities of product families. He balances the need for package changes and modifications with merchandising's need for customer identification with the package, eye appeal, durability, and economy.

The manager recommends basic policy to the merchandising groups and, upon approval, works within that policy. Should there be a need to go beyond this policy, he must first receive the approval of ____________ (depending on the company organization, the title of the person from whom he must get approval would be inserted here). Usually, disputes are arbitrated and new policies set in the office of the top official responsible for marketing.

The manager reports to ____________ (fill in according to the corporate structure). Reporting to him are the packaging art director (or graphics supervisor), the copy chief (and copy writers), and the packaging coordinators.

Dimensions

Include the number of salaried employees and the size of the department's estimated annual budget, which usually relates to the corporate packaging expenditures.

Principal Accountabilities

1. Assist in the development of the companywide packaging program to insure continuity in the corporate design.
2. Assist merchandising in determining specific approaches to package modification and design to insure that corporate objectives are met.
3. Coordinate the activities of specialized company sections and outside vendors to insure that the package projects are completed on schedule.
4. Direct the selection and supervise the activities of outside design firms to insure that costs are minimized.
5. Provide functional guidance to the legal, R&D, production, purchasing, and engineering personnel to insure that packaging programs are implemented.
6. Make recommendations, as requested, to marketing officers in subsidiaries to insure that the benefits of the packaging program are available to the entire corporation.
7. Direct the activities of the subordinates, select personnel, and train and develop staff members to insure continuity of the organization and the effort.

JOB SPECIFICATION FOR SUPERVISOR OF GRAPHICS (ART DIRECTOR)

Accountability Objective

Determine corporate packaging design needs and decide how best to meet them to insure optimum packaging for the corporation's products.

Nature and Scope

The supervisor (or director) is concerned with obtaining information on companywide packaging design requirements (primarily from merchandising management). He must possess and utilize a thorough knowledge of packaging techniques, printing methods, and the limitations imposed by the nature of the product and the dictates of corporate packaging requirements to guide and advise outside design studios. Attractiveness and function are the most important factors to be weighed in choosing acceptable designs. Since the supervisor works closely with all levels of management, good human relations skills are essential. He must also have good organizational abilities, imagination, and a comprehensive familiarity with the corporation's products. He reports to the manager of packaging design and must keep him fully informed on the status of packaging projects. Authorization must be obtained from the manager before a design proposal is implemented.

Dimensions

Salaried personnel count in the suborganization—that is, assistants, if any, and the approximate size of the director's packaging budget.

Principal Accountabilities

1. Determine corporate packaging design and size needs and advise management how to meet them in order to aid development of superior packaging.
2. Direct the activity of the various departments involved in packaging design to insure efficiency in that aspect of their work.
3. Schedule and expedite packaging projects to insure efficient and timely development of new or revised types of packages.
4. Furnish packaging design specifications and advise packaging development to enable it to efficiently solve technical packaging problems.
5. Keep abreast of packaging design innovations to insure that all packaging is competitive.
6. Prepare reports, as necessary, to achieve optimum co-

ordination of packaging design efforts and adequate dissemination of information.

JOB SPECIFICATION FOR PACKAGING DESIGN COORDINATOR

Accountability Objective

Coordinate corporate packaging development and provide a central source of information and advice on packaging requirements to insure timely availability of superior packaging.

Nature and Scope

The coordinator acts as liaison for the various corporate activities involved in packaging development—that is, production planning, purchasing, quality control, product management, packaging engineering, and so forth. This involves obtaining data on packaging requirements, conveying and explaining the data to the activities involved, and coordinating the work of the individual activities to produce a unified effort. His outside contacts are primarily with printers, engravers, and suppliers to instruct them on the details of corporate packaging requirements. He also institutes and directs cost studies to determine the practicality of projects and obtains final approval on all projects from each of the activities involved. He reports to the manager of packaging design and must keep him fully informed on the status of packaging projects.

Dimensions

Salaried personnel count in the suborganization—that is, assistants, if any, and the approximate size of the coordinator's annual packaging budget.

Principal Accountabilities

1. Coordinate packaging development activities of various corporate areas, such as product development, production planning, purchasing, packaging development, to insure efficient synchronization of work on packaging projects.
2. Determine and interpret packaging requirements in terms of size, material, product weight, and the like, to furnish packaging R&D with the information necessary for development of specifications.
3. Expedite packing projects to insure *timely* development of required packaging.
4. Assist purchasing in determining what will be practical artwork to insure early recognition of that which will be impractical.
5. Supply packaging information to production planning in order to develop cost studies.
6. Work with packaging engineering in determining availability of adequate equipment to insure that all packaging projects are feasible in terms of available or economically obtainable equipment.
7. Obtain approvals on packaging proposals to insure their meeting all the requirements of merchandising management and the other activities concerned with packaging development.
8. Act as an information center to insure adequate communication and adherence to schedules.

JOB SPECIFICATION FOR COPY CHIEF

Accountability Objective

Prepare all copy for the corporation's packaging materials, promotion pieces, inserts, sales brochures, and catalogs.

Nature and Scope

The copy chief's main activity is in packaging and associated areas, and his concern is with preparing copy for promotional needs other than advertising. He works closely with the merchandising managers, who provide the key information on the products under way, and recommends to them basic policy on copy approach. He also has a close relationship with the legal, advertising, and purchasing departments and with the packaging coordinators and art director.

The copy chief must have a feel for the corporation's products and packaging philosophy and a knowledge of the major state and Federal regulations on product claims and on warning requirements for potentially dangerous products. All copy must be cleared with the legal department as part of normal operating procedure. Since his position calls for close contact with people, he must have the ability to direct and motivate. He reports to the manager of packaging design.

Dimensions

Salaried personnel count in the suborganization—that is, assistants, if any, and the size of the copy chief's annual packaging budget.

Principal Accountabilities

1. Develop a corporate copy format for the overall packaging and sales effort.
2. Assist merchandising in determining specific and original copy approaches to the various types of products, thereby furthering corporate objectives.
3. Establish a system of copywriting and checking within the packaging framework of the corporation.
4. Direct the activities of subordinates, select personnel, and train and develop staff members to insure continuity of the organization and the effort.

Job descriptions are necessary to provide an orderly means of structuring positions and salaries, but no description ever really captures the job's flavor. And it must be remembered that these sample descriptions do not cover every tiny aspect of the job but merely establish the parameters of operation and control.

Contact Areas

Packaging design's main contacts are with the merchandising groups, the legal, purchasing, engineering, quality control, production-planning, and advertising departments, and with the sources. In addition, if it does not have its own copy section, it uses the corporate copy staff. All of these areas must become involved in the development of a package, so it might be helpful to summarize briefly each area's interrelationship with, and effect on, packaging design.

Merchandising. Since this group furnishes the initial concept of the project, in terms of sales area, market and product concept, relationship between product and package, and general image desired in the marketplace, it is, of course, vital to the design.

Legal. The legal office rules upon the necessary copy, such as advertising claims, premium offers, and expiration dates, for compliance with state and Federal laws referring to weight statements, ingredient statements, use claims, and other associated claims.

Purchasing. This department has a great influence on the design concept in terms of feasibility of reproduction, preparation of final artwork, costs, planning, and schedules, and generally has the final say on the design.

Engineering. The engineering unit assists design by advising on matters such as layout templates of various types of packages, machinery utilization and machinability problems, palletizing, plant handling costs, and other related subjects.

Copy. Packaging design departments without copy staffs of their own must rely on the corporate copy section, which has a direct-line relationship with the legal department. In most large merchandising houses there is a unit, generally referred to as the "trade practices department" and staffed by lawyers, that screens and approves all copy for state and Federal restrictions. It must be remembered that in writing copy, particularly in a house with many diversified items, there are myriad local regulations to contend with. The various states, and in some instances municipalities, have very stringent restrictions in certain areas. For example, California has strict laws on the labeling of bedding; New York City, through its fire department, is restrictive on items having a potential for fire, explosion, or poison. In companies without any copy staff whatsoever, the preparation of the copy is usually left to the outside design studio or to the personnel of the interested merchandising group. It is absolutely vital that this copy also be cleared with the legal department; to forget or purposely evade this should be grounds for dismissal. And it should be remembered that the ultimate blame for such an omission falls on the coordinator responsible and, lastly, on the packaging design manager.

Quality control. The printing (color) standards originally prepared by purchasing and the source and approved by the design group become the working tools of quality control. The department normally has no part in approving standards because, in general, its personnel are not technically equipped to evaluate the details of printing, but it uses the standards to compare incoming packaging shipments for adherence to the preselected standards.

Production planning. Since this department is responsible for producing the product to meet the marketing schedule, it is involved with the package from its inception. If the product is not new, production adapts its run to meet marketing's needs. If the product is new, and therefore often beset with the problems of breaking in machinery and personnel, there must be a tight coordination of the package-development scheduling at all times. Any scheduling change, whether in design, marketing, or purchasing, must be instantly transmitted to production. Conversely, any production delay must be instantly made known to all other concerned departments. This control of schedule is a never-ending problem.

Sources. If properly staffed, package-manufacturing sources can be one of the best aids to a design director and can make the task of preparing final artwork a very easy one. As a rule, a large producer of packaging materials has art personnel available for both creative design and mechanicals, but most design directors seem reluctant to use them. This is somewhat understandable in creative design but incomprehensible at the mechanical stage. The manufacturing source invariably produces better, more practical, more usable, more efficient, and less expensive mechanical art than a studio, and it does not suffer the studio's normal handicaps. If the source is not handling the art, it should nevertheless review the comprehensive, and any suggestions it offers for making production of the job easier should be considered. Generally, the source acts as a leavening agent for the design group and prevents the design of a package that cannot be properly produced. Purchasing's prerogative is to insist on this review before final art is created—it will save much time and money.

Advertising. Packaging design must keep close communications and scheduling ties with advertising, which must plan in advance the types of advertising to be used and which often uses the package design in its campaign. Depending upon the media, planning is often quite long-range and the cost is high. The advertising campaign is usually one of the major factors in the production and packaging-development schedule—there must be a product in the sale area when the campaign is in full swing or there will be a considerable consumer-relations problem. Nothing is more irritating to a customer than to have his desires whetted by a good advertising campaign and then not be able to buy the product.

These then are the key contact areas for package design. Each has an impact on design and upon scheduling. The coordinators work with the merchandise groups to develop a detailed schedule for every function involved in producing a package on time, but more important, they must keep everyone concerned on the schedule, which requires constant communication with all departments and demands that necessary decisions be made promptly.

Project Priorities

The question of what gets done first is not a simple one because almost every party in the scheme of developing packages gets involved. The manager of packaging design usually receives (or should seek) guidance from the leaders of merchandising as to what projects should have priority, but complications set in when purchasing gets into the act with requests from packaging R&D for cost-reduction projects. Actively, project priorities for R&D are established by the needs of the merchandising groups, but they are affected by such activities as planned CRD tests, test markets, advertising coverage, product-development schedules, and plant production capabilities; therefore, the only hope for an orderly schedule of priorities is to set realistic dates for accomplishment of the various stages. So far, so good. But now the weakness of the normal packaging-development procedure becomes apparent. Anyone can set up a good schedule, but only strong, determined people can make it work, and if all its facets are not constantly watched, checked, and acted upon, the schedule falls apart.

The packaging design manager, more than any other person, must control the schedule and insist on, and get compliance for, the dates. The coordinators are the keys here—they are the legmen, the bird dogs if you like, who must keep their eyes on a million details at once. They are responsible for either smooth accomplishment or chaos. The manager must educate, train, and check up on these men to insure that they do the job properly. In most cases where coordinators are not functioning properly, the fault lies with the manager's lack of administrative skills.

Design Sources

Design work must be differentiated from mechanical or final artwork. Design is creative, a flow of concepts usually evidenced by sketches at the early stages, package mock-ups or comprehensives at the intermediate stage, and lastly, final or mechanical art. The description is simple, but the task is not. Final art preparation and color selection often determine the fate of the package and the product in the marketplace. The technical ramifications of color choice, layout, color laydown, quality of illustration, and other vital considerations are beyond the scope of this book, but they are vital areas for design and, most of all, for purchasing.

There are several ways of executing designs, but essentially the design manager has three means of creation: the design or advertising agency, in-house design, or source design facilities. Most art directors prefer using the studios, which we discussed rather extensively earlier in the book, but experience indicates that a majority of companies also go through the phase of having their designs executed by artists on their own payroll. This approach, however, is eventually abandoned in favor of outside facilities. The reasons for this seem to stem from the difficulty of finding the right creative people, keeping them stimulated but yet controlled, and from the gradual stagnation that develops from working in a limited field.

Arguments advanced by design directors to explain their apparent reluctance in using the often excellent design facilities of sources supplying packaging for the corporation are never very strong, but they generally evolve from a feeling that source people are creatively restrained by the limits of their manufacturing capabilities. To an extent, this is probably true, but at the same time there is a practicality in having the source design what it will actually manufacture. Often, an outside studio designs a package that, because of technical problems, cannot be faithfully reproduced or manufactured within the cost limitations. But no matter which method of design the director chooses, he can get the desired result *if* he knows what he wants, relays the instructions clearly, and is willing to work with the people involved.

This controversy over use of source art facilities has gone on for a long time between purchasing and design; purchasing usually concedes in the area of creative design but still fights the battle of mechanical art. Undoubtedly, the manufacturing source can prepare mechanicals better, quicker, and cheaper than a studio. There are no pluses for the studio. The source knows how art should be prepared for *its* equipment and does so accordingly. But a studio, on the other hand, has no knowledge of the equipment that will run the job, the potential production trouble spots, and, in general, is unaware of the type or method of printing to be used (art must be prepared in slightly different ways, depending on the printing method). Artwork prepared by a studio usually must be reworked by the manufacturing source. Preferably, the procedures for source production of the final artwork should be established slowly—step by step. Nothing succeeds like success. Purchasing has a perfect right to set up the technique on any given job. Progress should be made slowly, and eventually a working "way of life" will evolve with specifically selected packaging sources.

The packaging design manager occupies a pivotal position in corporate packaging management. He acts as an overall controller or coordinator of schedules, approvals, and project priorities, and, in general, is charged with making everything run smoothly and on schedule.

Chapter VI

Packaging Procurement

THE THIRD MEMBER of packaging's big three, packing procurement, is basically a profit-making operation that can have a great effect on the profit margins of the separate marketing groups and on the overall profit picture of the corporation. Buying packaging is not an ordinary purchasing function. The concepts of operation, particularly if there is multiplant servicing, are considerably more complicated than those of other procurement areas. Coupled with this is the fact that packaging purchase is virtually a custom business—in most instances, there is no such thing as a "standard package." Each run, each deal has highly distinctive characteristics.

Purchasing also has a heavy communications burden, a fact that has been instrumental in changing the techniques of packaging buying and has forced the use of every available system of reducing detailed work to leave the buyer free for creative activities. Not too many years ago buying was buying, and it could easily become routine.

Organizational Structure

As a rule, the packaging procurement function is part of the general purchasing department, which buys everything needed for running the business. Usually, packaging procurement operates as a separate unit reporting to a manager who, in turn, reports to a vice president or director. This separate identity concept normally extends to the plants, some of whose personnel are specifically assigned to the task of packaging control and are responsible for ordering all the plant's package supplies. The department's structure will probably closely resemble that of R&D. The reason for this is that the major approaches to its organization are basically the same : industry approach, product approach, or a hybrid of the two. There are, however, enough differences between the two departmental setups to warrant a brief reappraisal of the approaches from purchasing's viewpoint.

Organization by industry or type of packaging. Under the industry approach, agents or buyers are assigned to designated areas of the packaging industry, just as R&D assigns technologists. In general merchandising houses this is the only way to operate. Years ago, one of the major houses experimented by assigning a packaging buyer to a group of merchandising departments; he was responsible for purchasing all their packaging needs. But this system was found to be less efficient than the industry-orientated approach for a variety of reasons. First, if the buyer has to handle more than two or three packaging types his knowledge usually gets diluted. In addition, problems crop up over duplication of sources, production priorities (when more than one buyer is using the same source), source development procedures, and source relationships in general. However, the approach works quite well for R&D because of its need for a variety of materials and the fact that it has no pressing volume problems, which purchasing has. The advantage of industry orientation, however, is the same for both purchasing and R&D: The buyer can become an expert in his field. But there are additional benefits in the fact that communications between buyer and industry are simplified, the buyer can concentrate on source development since he works with relatively few sources, and he can do more volume buying. One of the system's disadvantages, which R&D also shares, is that buyers should be rotated every few years. Jobs can become routine, and a change introduces a fresh viewpoint—but it also means training new men almost constantly. The major disadvantage, however, centers around communications between the merchandisers and the buyer. The contact problem is particularly difficult for the merchandiser since he generally needs several components to make up one package. The burden of contact lies with him, and it can be quite a heavy one from a planning and control standpoint.

This problem of multiplicity of packaging buyers working on a given project can be eased somewhat by assigning a project leader. But if this is done, one primary component is to have a project control system. The easiest method is to have all projects assigned by the department manager, who is the logical man to head the control system, because he is in a centralized position and therefore can balance workloads and do the necessary follow-up. He also has the authority to get action. The organization chart would resemble Exhibit 6.

Organization by product. This organizational effort is gaining some popularity with companies that have very distinct delineations between products and where each merchandising group handles a specific type of product or group of products. The idea is to assign a buying team, headed by a buyer, to each category. (See Exhibit 7; the purchasing department's buyer is akin to R&D's project leader.) In theory, each team buys all the packaging materials it needs, so the centralization of effort and good communication are definite benefits. But if practiced in its purest form, the idea has some grave problems. Critics contend that it puts an extra burden on the salesman calling on the account because of the additional people he must contact. This is no doubt true, but it also grants him a broader exposure to the company.

Exhibit 6

INDUSTRY APPROACH (FOOD INDUSTRY)

MANAGER
BUYER Labels, Tags, Wrappers
BUYER Plastics
BUYER Folding cartons
BUYER Corrugated containers, Multiwall bags
BUYER Cans, Spiral-wound tubes
BUYER Small bags, Rollstock and balers
BUYER Glass

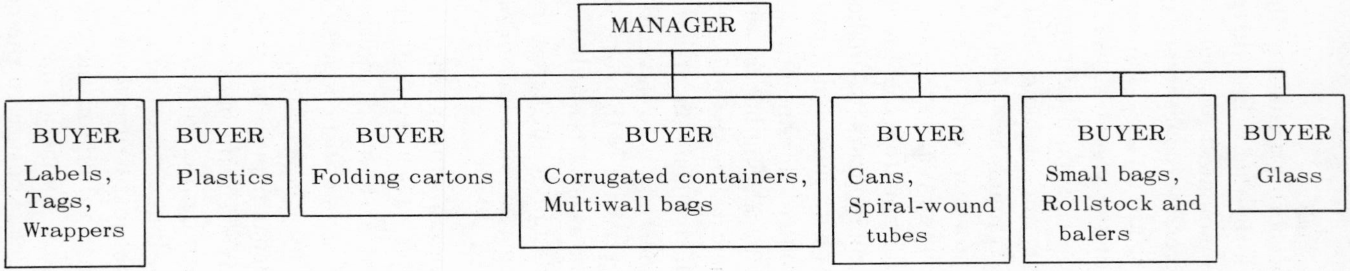

Exhibit 7

MERCHANDISING APPROACH (FOOD INDUSTRY)

MANAGER
BUYER Cereals team
BUYER Pet foods team
BUYER Cake mixes team
BUYER Waxes and detergents team
BUYER Desserts team

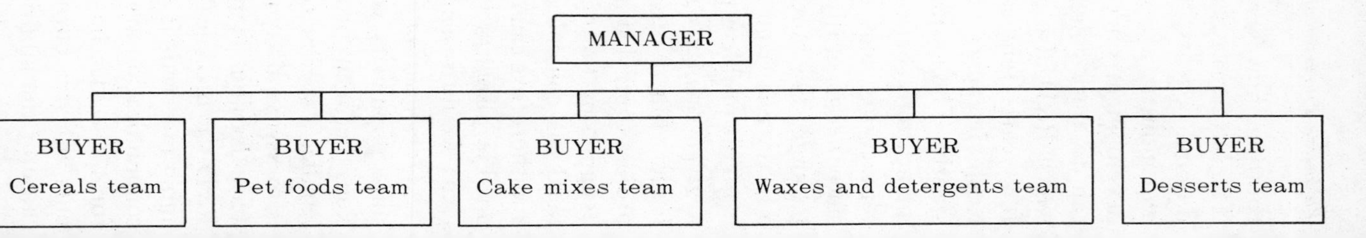

There is a more serious fault with this technique, however, which is that the buying team may not be able to indulge in volume buying. And it should also be pointed out that this approach, again in its purest form, hampers the team's ability to maintain large quantities of such items as containers, cans, and glass bottles. Therefore, as with R&D, the trend has been toward a compromise.

Hybrid approach. If the corporation is a big user of many kinds of packaging materials, which is usually the case, a hybrid organization might be the solution. This will, however, require some realignment in communications to effect a tie-in with merchandising. Generally, the best way to do this is to standardize the instructions and make sure that everyone concerned knows what they are. Intrahouse communications is a subject suitable for extended corporate study because decision making and communications are the major problems in almost every industry today.

Under the hybrid system, the department would be structured in much the same way as previously outlined, but there would be fewer teams and more industry lines (see Exhibit 8). This particular organizational technique is adaptable and expandable to fit into almost any kind of corporation producing an item for the retail markets. It is particularly good for food companies because it clearly pigeonholes the product groups. But it is not as easily adaptable to general merchandising houses because of the multiplicity of their items. Experience indicates that these establishments should follow the industry approach, provided that their buying staffs are given enough coordinators to help ease the communications burden. Exhibit 9 shows how a typical team's responsibilities would break down if the hybrid approach were used. The variety of items illustrates the various industries involved and the scope of the buyers' interests. There is no doubt that in some respects their jobs are more difficult because of the range of knowledge and contact required, but this disadvantage is offset by the easing of the communications problem and the simplifying of procurement planning or the various related package components. It is also easier for a buyer to chart

Exhibit 8

Hybrid Approach

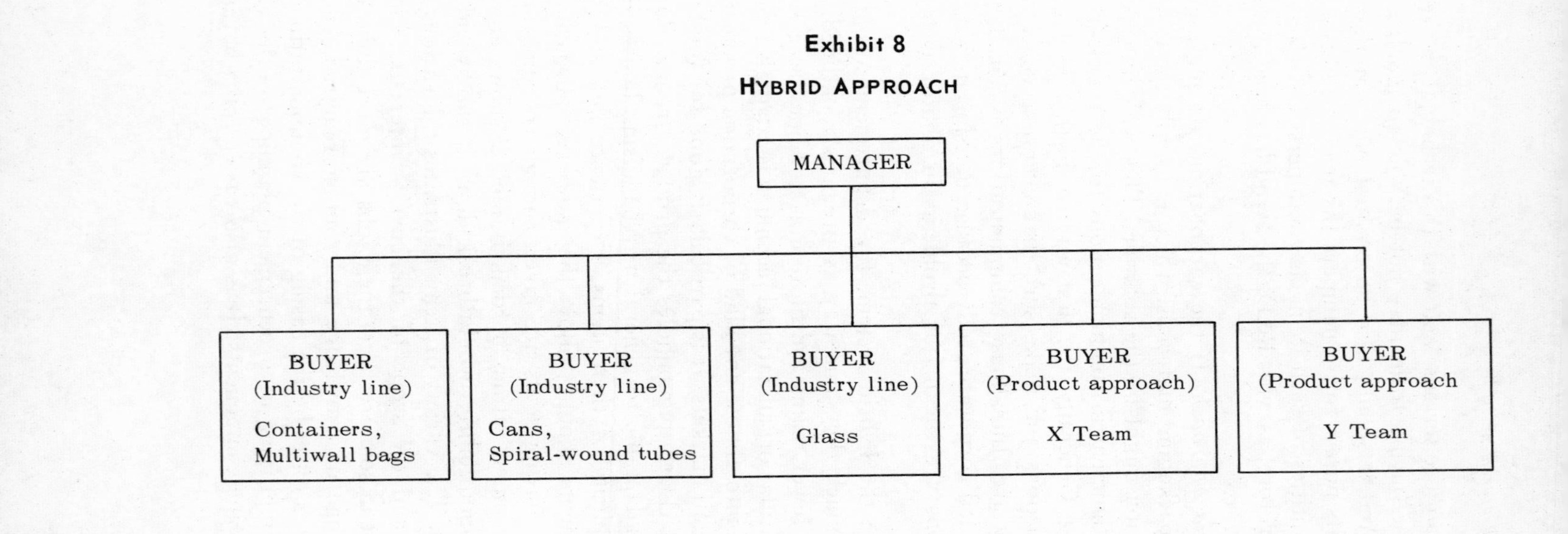

Exhibit 9

Typical Hybrid Team's Responsibilities

READY-TO-EAT AND INSTANT CEREALS TEAM

(Items to be Purchased)

Folding cartons
Liners, glassine and foil, or the like
Pouch paper
Inserts
Specialty boards
Overwrap film

the package's current position because he controls all of its elements.

Buyers operating under the hybrid system report, after a period of adjustment, that they become much more aware of the corporate marketing objectives. This, in turn, has a tendency to instill in them a greater sense of participation in the growth of the product lines. In the specialized world of business it is very easy to develop a limited horizon of interests and to fail to feel that procurement is really part of the entire corporate effort. The concept of getting the package-buying personnel more involved in marketing is relatively new, mainly because purchasing has traditionally operated as a separate entity, but it is one of the excellent approaches used today to increase packaging procurement's participation in marketing.

Purchasing Services

Purchasing services, a relatively new addition to the purchasing department, performs a variety of functions designed to relieve the buying units of some of the detail work. It should also serve as a communications medium. The following is an outline of its basic responsibilities. The scope of activities is easily expandable as the range of the purchasing department's interest grows.

DEPARTMENT CONCEPT

This unit is set up as a central administrative agency for management information, for coordination of projects during the early development stages prior to individual buyer participation, for control of the various reporting systems utilized by the purchasing department, and for assembly of technical and administrative data necessary for planning.

Specific Areas of Responsibility

The supervisor of purchasing services provides assistance in the following areas:

1. Purchasing research (systems). This is a broad function covering special assignments, control of systems development, and investigation of ways to reduce administrative burdens and costs.
2. Management information coordination. Purchasing services is the primary contact for those charged with development of the corporate management information system. The actual development and operation of any mechanical systems are the responsibilities of the corporate systems and data processing departments.
3. Corporate project coordination. The supervisor provides a contact point for coordination of information requests from other departments during the planning stages. The basic purpose is to establish master files for all packaging projects and any other projects, such as long-range cost analysis and machinery studies, that are in the works. The packaging project files are based on the development status/priority list issued by packaging design or R&D. All information, letters, specifications, and the like pertaining to a project must be kept in these master files. This does not preclude each buyer's keeping his own files, but it is his responsibility to see that copies of any relevant communications are sent to purchasing services for inclusion in the master files.

4. Management reports. This office preserves information and is prepared to submit reports. Reports are both scheduled and on an as-required-by-management basis.
5. Obsolete-inventory control. The department is responsible for the operation of an obsolete-inventory system that identifies obsolete items and decides how they are to be disposed.
6. Cost-study control. All cost studies are controlled by purchasing services in order to provide production with a central point of contact in this joint operation.
7. Central files. The central file includes departmental reports, procedures, manuals, buying guides, bills of materials, and any other data needed by management. A technical-data file is maintained as a departmental reference source, and buyers are responsible for forwarding data to it.
8. The supervisor is responsible for preparing and implementing departmental buying budgets and operational procedures and manuals, for controlling overall inventories—both at plants and sources—for advising buyers on inventory control procedures, and for acting as controls and systems adviser to related personnel at the plants.

Packaging Procurement Responsibilities

Regardless of how the corporation chooses to organize the packaging procurement operation, the goals of the group do not vary to any great degree. Sound purchasing is based upon buying materials of the proper quality, in the proper quantity, from the proper source, at the proper price, with delivery at the proper time. But although packaging purchasing adheres to these basic rules, it goes much deeper because of the pivotal position it holds within the corporation, which depends on packaging to get its product to the marketplace. The department's responsibilities, when broken down and studied, are great indeed. Not only can an inefficient purchasing group disrupt production, but it can be a source of considerable profit loss for the company, so its fundamental responsibilities should be clearly spelled out. The following are its main purchasing obligations.

Planning. This means preparation of both long- and short-range plans, which, at best, involves a one-year and a three-year plan. These should cover such items as cost-cutting projects; implementation of EDP systems; reduction of paperwork; a counseling system with the merchandise groups; long-range organizational concepts; personnel planning to cover training, promotions, acquisition of new personnel, and improvement of individual communications, consultations, and development; and profit contribution.

Supply (logistics). Package purchasing is responsible for seeing that the necessary packaging supplies are available to the manufacturing units as needed. This requires the development of a varied, yet simple, communications system that ties together purchasing, the plans, and the suppliers. The department must assume the burdens of inventory control, training of plant ordering personnel, reduction of losses and write-offs, and control of all related expenditures.

Quality. Only purchasing can control the production of quality packaging that meets the corporate guidelines. And since quality printing comes from quality artwork, purchasing must, despite resistance, demand the kind of art it needs to do the job correctly.

Source selection. This is probably one of the major keys to the success of the purchasing effort. Source development, not just selection, is a major task of the buyer, but a source must first be selected and certain basic rules apply, particularly for packaging sources. They should be chosen with an eye toward:

- Ability to provide packaging research and development assistance to the company's R&D people.
- Geographic location, which plays an important role because of the weight of most materials and the desire to reduce freight costs by not having to ship excessive distances.

- Equipment and technological facilities to produce quality work as required.
- Service for the packaging buyer, which goes beyond the normal concept of getting the goods on time. Package purchasing requires many extra items, proofs, inventory reports, breaks in schedule (in emergency situations), samples, artwork, ink-maker and plate-maker consultations, and a host of other things that tend to make the buyer's job easier.
- Pricing. In packaging, as elsewhere, you get what you pay for. Quality costs money.
- Attitude. This is a key factor because every relationship has its moments of strain. The way the source handles the corporate vagaries is important. Attention to detail and compliance with requests for information and assistance are vital factors for long-range buyer-source cooperation.

Each purchasing man undoubtedly has his own list of requirements for selecting a source, but in today's industry it is a good idea for him to select carefully with the hope of developing a team effort and a relationship that will last beyond his tenure.

Coordination. It seems that as the importance of packaging grows in a corporation, the coordinating task of purchasing increases. This is an important responsibility for the department because it appears that this area will be the instigating force behind future radical changes in the way the company handles its packaging. And these changes are unavoidable.

Profit contribution. Purchasing is a profit-making operation. A few years ago this was not a popular idea; today it is an accepted fact of business life. In a corporation with a heavy packaging orientation, purchasing's chances for contributing to the profit picture are enormous. As a rule, the department operates against a predicted or estimated standard cost, and variations from a standard can be a measure of performance because cost savings related to this factor are, in truth, increased profits. In the food industry where the case cost, which is the total cost of everything needed to produce a case of goods, is the basic criterion, any cost savings are reflected in a vivid fashion. The key to profit success is careful planning in the selection of cost-saving projects, close follow-up, and good documentation. In setting the projects, the buyers are the prime forces because of their intimate knowledge of the items involved. Consulting with the buyers on a regular basis, reviewing progress, stimulating their thinking, and increasing motivation for cost savings will pay great dividends to the corporation.

Personnel Requirements

In many segments of the purchasing field a growing feeling has developed that the packaging buyer is unique among all buyers. Packaging, an expanding industry closely allied with marketing and merchandising, requires a special type of man as buyer. The know-how needed for this position has always been staggering, but packaging's growing association with marketing necessitates the buyer's having a wider range of knowledge at his command, with consequently higher educational requirements. To be a good technician is no longer enough. Today's packaging buyer must be a multifaceted leader, with not only authoritative technical mastery of his areas of packaging control, but also the ability to comprehend the languages and functions of marketing, design, and advertising. He must be able to express himself well, lead as required, and motivate others.

Packaging procurement in a heavily package-oriented corporation is big business and must be recognized for its overall contribution to the firm. In order for it to operate effectively at a profitable and efficient level, the personnel must be the best available, and sufficient salaries and other incentives should be provided to this end. But to do this, the job must be properly rated.

The key to securing proper job evaluations and accordingly suitable ratings and salary scales is the job guide or job specification. But evaluations are generally made up by a committee that is unaware of the position's complexities. Awareness of this hazard of committee ignorance is the first step in winning the battle of job and salary ratings. Poorly written or carelessly constructed evaluations can be the opening shot in a campaign that could reduce purchasing to a mundane role. It is very important that the key rating areas be covered in the job specification in explicit language that will grasp the attention of the evaluation committee.

First, however, it is necessary to know how to prepare an intelligent and logical position concept, which is the basis for the job specification and differs from it in emphasis, detail, and language. These discrepancies become apparent when pinning down duties in the specification. The importance of language cannot be overemphasized—it might make the difference between excellence and mediocrity in acquisition of personnel.

The composition of job specifications must be approached with intelligence and knowledge of what is required to stir the imagination of the evaluation committee. The key is to discover what type of system is being used, to study it, and to understand it. Then you are ready to combat the system.

Most systems rate requirements, such as know-how, in terms of background, education, and experience needed for the various job levels, coupled with some sort of evaluation as to the breadth of the job in the corporate structure. Also, there is usually a rating area for the impact of the human relations skills required, generally in terms of the job's effect on influencing people. Problem solving invariably occurs as a rating criterion, usually divided into two concepts: depth, in terms of direction or guidance required and degree of freedom of action and decision allowed; and breadth, in terms of the types of situations that the job-holder is expected to handle. For example, a situation statement might read: "Variable situations requiring analytical, interpretive, evaluative and/or constructive thinking."

Accountability is generally reckoned by the amount of policy guidance the job requires and the budget responsibilities of the function. Impact or effect of the job on other units is often a consideration, particularly the amount of participation with others inside or outside of the department that is required before taking action.

To demonstrate the difference in approach between the position concept and the job specification, compare the language, scope, and technique of the following two examples for the manager of packaging materials procurement. The first example, which follows, presents a position concept that might be prepared when a new position is created or an old one reworked.

MANAGER OF PACKAGING MATERIALS PROCUREMENT (POSITION CONCEPT)

Basic Guide

Plans, directs, and controls policies and procedures for the purchase of packaging materials for all product lines; provides functional guidance to the plants in matters relating to the acquisition and control of packaging materials.

Major Responsibilities

1. Directs the development of, and recommends for approval, master contract, vendor selection, and vendor evaluation policies and plans.
2. Directs the development of, and recommends for approval, packaging materials control policies and procedures (including appropriate IBM procedures) for the checking of vendor inventories against master contracts, for plant inventories, and for the efficient administration of the packaging materials group's activities.
3. Directs the development of, and recommends for approval, annual cost-reduction projects, packaging and methods improvement plans, and goals for the packaging materials group; develops performance-reporting procedures for measuring progress against these goals.
4. Directs the control of packaging materials inventories held by vendors against master contracts to minimize inventory accumulation and obsolete-materials write-offs

while maintaining adequate stock to satisfy plant demand.

5. Directs assigned buyers to act as primary contact points for suppliers and potential suppliers of packaging materials; sees that buyers establish liaison between vendors and other departments wherever such direct contact is in the corporation's best interests.
6. Approves the selection of all packaging materials vendors as recommended by the various packaging materials buyers under the master contract program.
7. Reviews and approves all claims against vendors stemming from unsatisfactory materials, untimely delivery, or similar vendor failures.
8. Provides functional guidance to the plants on matters relating to the acquisition, control, storage, and use of packaging materials.
9. Directs the development of plans for the sale of scrap, obsolete, and surplus packaging materials.
10. Works with merchandising to coordinate purchasing's functions in the development of new packages, coordination of deals, and similar situations requiring package changes or expert printing advice.
11. Works with the premiums manager to coordinate purchasing's duties in the scheduling and development of in-package promotions, including coupons, premiums, and the like; assists as needed in the search for, and selection of, vendors for premiums.
12. Sees that liaison is provided between purchasing and the production, engineering, and other departments in the establishment of new packages. This includes planning initial purchase quantities and setting up production dates.
13. Works with quality control to insure that package material standards are consistently met by all vendors.
14. Works with R&D on adequate and practical standards that can be met by packaging material vendors.

15. In keeping with corporate policy, takes charge of the department's normal personnel functions, including personnel training and development, evaluation, recommendation of salary increases, terminations, new employees, and the like.
16. Keeps abreast of trends and developments in packaging materials and makes suggestions for changes in the purchasing approach to take advantage of these developments.

Organizational Relationships

The manager reports to the vice president of purchasing and supervises all buyers and assistant buyers assigned to his department.

MANAGER OF PACKAGING MATERIALS PROCUREMENT (JOB SPECIFICATION)

Accountability Objective

Plans, directs, and controls purchasing and adequate inventory of packaging materials to insure quality, optimum merchandising advantage, and quantity, thereby meeting production requirements on time and achieving the best price advantage.

Nature and Scope

The procurement of packaging materials requires that the manager have a thorough knowledge of vendors, materials, printing, pricing, supply systems, and merchandising of packaged products. He protects the company against unnecessary or unwise expenditures through accurate and efficient purchasing. Each new development in packaging materials offers a potential merchandising advantage, and it is his responsibility to be expert in all these developments so as to protect and improve the company's marketing effectiveness.

The manager negotiates mill contracts and national packaging contracts to guarantee timely availability of quantity and quality materials to meet production requirements. It is necessary to plan and closely control contracted inventories at the plants to insure this. The manager seeks new sources at plants that have any available capacity, including development of master contracts, vendor selection, and evaluation procedures.

The manager analyzes cost estimates of any wholly owned packaging facilities, such as papermills, for various grades of packaging materials to determine which products can be produced at advantage in the plants. He organizes, motivates, and directs packaging materials staff on personnel matters, appraises their performance against predetermined goals, and assigns buyers to act as primary contacts for existing and potential suppliers. He develops packaging materials purchase plans and policies, including objectives and controls for cost reduction, and he acts as an adviser to all departments in solving problems related to the packaging area.

Providing functional guidance to plants on matters relating to acquisition, control, storage, and the use of packaging material is part of his job, as is assisting plants in handling any problems that may arise. He maintains a cooperative and productive working relationship between his people and those in the departments with which they work (that is, marketing services, product management, production, engineering, and so forth). He directs the development of plans for the sale and disposal of scrap, obsolete, and surplus packaging materials.

Dimensions

Include the number of salaried employees and the size of the department's annual budget, which relates to the corporate packaging expenditures.

Principal Accountabilities

1. Plans, directs, and controls purchasing and inventory of

packaging materials to assure timely availability in quantity and quality at an advantageous price.

2. Provides expert knowledge of sources for glass, twine, board, paper, corrugated box, plastics (rigid and flexible), and all other packaging materials to insure optimum merchandising advantage, timely delivery, the required quantity and quality, and minimum costs.
3. Organizes, motivates, develops, and measures performance of staff members and plans for continuity of the packaging materials organization.
4. Conducts long- and short-range planning for cost-reduction programs.
5. Plans for maximum utilization of EDP in handling the department's work.
6. Develops the packaging buyers' marketing participation.
7. Plans for, acquires, and trains personnel needed for ordering packaging materials at the plants.
8. Develops control systems for all aspects of packaging purchasing.

When reviewing the job specification, note the use of certain key words and phrases designed to give the evaluation committee a peg upon which to hang a favorable decision. Check the following key rating areas against the job specification:

1. *Knowledge*. This is covered in the first sentence of the opening paragraph under Nature and Scope.
2. *Problem solving*. This is a major consideration in the evaluation of the purchasing job. The important aspect is the easy recognition of the wide scope of problem solving involved.
3. *Accountability*. This is specifically covered in the section on personnel, budget, and purchases. Contact with other departments in the company is covered under the working-relationship provisions of the write-up.

These are the three keys to keep in mind when making up either a position concept or a job specification. It is much easier in the typical corporation to achieve maximum results with the evaluation committee the first time around than to have to submit a revised document. Work carefully, organize the functions as clearly and efficiently as possible, and spend the time necessary to do the job right.

In view of the importance of a proper approach to a specification, a few representative samples follow: buyer of packaging materials (industry approach); buyer's assistant (administrative clerk); supervisor of purchasing services; and buyer of packaging materials (product approach).

BUYER OF PACKAGING MATERIAL (INDUSTRY APPROACH)

Accountability Objective

Is responsible for purchasing assigned items and services within authorized limits, at the lowest price, and on the most favorable terms consistent with established quality, quantity, delivery specification, and vendor service.

Nature and Scope

- Buys authorized commodities according to purchasing operations and production needs.
- Participates in the preparation of and personally executes buying plans.
- Initiates and formulates master contracts for packaging and materials suppliers; constantly develops new sources of supply.
- Negotiates terms of contracts and expedites purchase orders.
- Selects sources and contacts; receives and confers with salesmen, manufacturers' representatives, service execu-

tives, plant managers, and the like to facilitate the most economical purchases.

- Approves invoices and follows transactions through to their end.
- Recommends changes in physical specifications, formulas, delivery schedules, or vendor services to permit more economical purchases.
- Maintains buying information sheets for use by the plants in ordering supplies.
- Prepares reports and records on all commodity transactions and incorporates them into EDP files.
- Establishes quality control standards for suppliers; polices the production of company supplies and materials.
- Visits suppliers' plants on a regular schedule.
- Provides suppliers with detailed specifications.
- Polices suppliers to insure compliance with specifications.
- Provides technical guidance to suppliers on questions of production costs and methods.
- Works with packaging design and quality control to develop and set standards of approval.
- Works closely with packaging development, design, advertising, production, and merchandising groups in connection with artwork, colors, production schedules, and quality reproduction.
- Conducts purchasing research programs, collaborating and gathering information on market availability and developments in commodity lines.
- Studies and keeps abreast of costs, market conditions, new items and equipment, and quality of work available in the industry.
- Maintains records and prepares reports as requested by management.
- Visits company plants periodically to establish a working agreement between plant and suppliers and the home office.
- Expedites shipments to plants when necessary.

- Provides technical assistance to plant purchasing personnel.
- Develops and maintains favorable relationships with approved sources of supply for items and services within assigned purchasing classifications.
- Keeps suppliers informed of anticipated requirements.

Principal Accountabilities

1. Purchases quality packaging materials needed for use by the company.
2. Develops master contracts and ordering systems for the using plants to obtain the needed materials.
3. Develops and maintains relationship with suppliers, R&D, packaging design, marketing, and other interested departments.
4. Keeps buyers' guides up to date and maintains a steady flow of communications.
5. Keeps abreast of market conditions.
6. Keeps management advised of all new developments that might prove beneficial to the company effort.

BUYER'S ASSISTANT (ADMINISTRATIVE CLERK)

Accountability Objective

Provides buyer with general administrative assistance to enable him to devote more time to nonroutine buying functions.

Nature and Scope

This postition relieves the buyer and assistant buyers of the routine and semiroutine activities of the purchasing function. The buyer's assistant is concerned with such things as placing orders, preparing contracts and processing invoices, checking shipments and the plants' source inventories, pricing inventories, and preparing buyers' guides. Although these activities are basically standardized, errors on the part of the assistant could have serious consequences. Preparing contracts involves verification of specifications, costs, and so forth (standard contract forms are used). The assistant is also responsible for actually placing the contracts (often by phone), and for such things as shipping dates, routings, and addresses. He reviews plant and inventory reports, alerts the buyer to problem areas, adjusts contract inventories against shipments, and supplies the management information system (MIS) group with necessary data—that is, such information as prices, usage factors, supply sources, and contract numbers. Another of his concerns is to proofread press sheets against original artwork and to consult with the manager on a variety of buying matters when the buyer is not available. This position requires a high school education, previous work experience, and a capacity for accuracy and speed in detail work.

Principal Accountabilities

1. Prepares contracts, assuring that specifications, costs, and the like are correct.
2. Processes invoices, making all necessary corrections before stamping for payment and forwarding to buyer for approval.
3. Places contracts, verifies information contained on the contracts, and advises plants as necessary.
4. Prices inventories, reviews plant and inventory reports, and alerts buyer to problem areas.
5. Supplies information to the MIS group as requested.
6. Handles phone calls, relays buyer's instructions or inquiries, and gathers data requested by him.
7. Prepares and updates buyers' guides (as directed) to provide guidelines for plant buyers.
8. Acts as a backup for buyers, consulting with the manager on matters when the buyer or assistant buyer is not available.

SUPERVISOR OF PURCHASING SERVICES

Accountability Objective

Manages the administration of the corporate purchasing function to assure timely planning and services for effective performance of both long- and short-range corporate projects.

Nature and Scope

The primary concern of the supervisor is to provide planning and administrative services to achieve timely coordination of purchasing projects and control of expenditures. He analyzes departmental activities and works closely with systems personnel to assist in the development of systems and techniques that will maximize the effectiveness and efficiency of the overall purchasing department. A related activity involves providing functional guidance to plants about proposed plant-level information centers (in reference to such things as prices, suppliers, and items). Another important activity requires acting as the administrative coordinator of various purchasing projects during the early stages, prior to final buyer participation. The supervisor is frequently called on to price physical inventories and interplant packaging materials transfers and to handle administrative aspects of "deal packout"* coordination. He also develops information for packaging, administrative, and technical files. Each of these files must contain all current and historical data, reports, and manuals that might prove useful to management or buyers. Additional activities of this position include overseeing the operation of the obsolete-inventory system, carrying out research on purchasing techniques and administration, and analyzing office workflow, layout, equipment, and furniture.

Since the supervisor of purchasing services works closely with systems personnel, he must have a knowledge of systems analysis

*"Deal packout" is a term used to describe the packing of bargains. For example, a 4-cents-off deal on dog food might require new labels and can lids announcing the bargain. Since the packing will be done at perhaps four different plants, the trick is to balance the supplies.

and computer usage as well as of statistical analysis and purchasing techniques, source development, suppliers' manufacturing processes, and paperwork systems. He works within a broad policy framework established by the vice president. Major out-of-the-ordinary problems are discussed with superiors before action is taken (especially problems that may affect other departments).

Dimensions

Include the number of salaried personnel in the suborganization and the approximate annual expenditures for machinery, equipment, and packaging materials.

Principal Accountabilities

1. Provides planning and administrative services to insure proper coordination of purchasing projects and control of expenditures.
2. Administers the purchasing department segment of the established management information system to insure timely availability of accurate information regarding purchasing department activities.
3. Prepares special reports for management, as requested, to supply information on any specified area of packaging.
4. Coordinates and controls all cost studies conducted by purchasing to provide production with a central point of contact in this area of joint operation.
5. Develops material for inclusion in purchasing's section of the corporate plan.
6. Maintains the inventory control system and reporting activities to insure maintenance of up-to-date information on the quantity and location of all in-stock items.
7. Participates in the operation of a value analysis program and acts as head of the value analysis committee to insure that the purchasing department's procedural/administrative changes are justified from a financial viewpoint.

8. Prepares departmental procedural manuals and maintains technical and administrative files to provide central sources of information in many areas (exclusive of those procedures prepared by data processing and systems personnel).
9. Maintains and controls a numbering system and master information on materials.
10. Maintains and controls a packaging bills-of-materials system.
11. Analyzes and issues reports on the corporate plan and departmental budget variations.

BUYER OF PACKAGING MATERIAL (PRODUCT APPROACH)

Accountability Objective

Purchases all packaging materials required to maintain efficient production of the company's ready-to-eat (RTE) and instant cereals.

Nature and Scope

This buyer functions as a purchasing "generalist" in the area of RTE and instant cereals packaging materials. He purchases all materials required for these two company products; cartons, liners, inserts, and pouch paper are typical. Planning purchases carefully to maintain an optimum corporate inventory in terms of both short- and long-term requirements is a must, as are expediting purchases and coordinating the orders to maximize transportation efficiency. The fact that his efforts are focused on specific product lines makes it easier for the buyer to coordinate purchasing activities with those of product management. He also works closely with the marketing managers to develop precise and detailed timetables for purchasing activities. These timetables must take into account promotions, new-product introductions, and plans for expanded distribution. Since the position deals exclusively with nonstandard items, considerable ingenuity is often required in locating suitable sources and in aiding them in creating items that meet specifications. The buyer is free to purchase required items without consultation, providing that they meet specifications. Changes in suppliers for major items or changes in established purchasing procedure are discussed with superiors. He constantly strives to improve purchasing methods (systems) and to find or devise materials that meet company needs more effectively than those currently used. He works closely with plant personnel, suppliers, product management, packaging R&D, design, and production planning.

Extensive knowledge of graphics, materials, plant operations, and company products and equipment is essential; five to ten years of purchasing-related activity should provide adequate background. And the position requires foresight; the effects of current purchasing activities and prospective market conditions on future supply requirements must be anticipated. High personal integrity and excellent human relations skills are mandatory, as is a capacity for original thought.

This buyer reports directly to the manager of packaging materials procurement. Reporting to him are an assistant buyer and a buyer's assistant.

Dimensions

These are determined by the approximate annual expenditures of this position (fill in amount according to the situation) and the number of salaried employees in the suborganization (varies by company).

Principal Accountabilities

1. Purchases and helps select quality and economical materials to insure maintenance of efficient production of RTE and instant cereal packages.
2. Evaluates suppliers and their materials to insure prompt service and full value for the purchase dollar.

3. Keeps abreast of new packaging products and improvements in order to advise plants and management of desirable innovations.
4. Maintains excellent relationships with supplier, plant, and merchandising personnel to achieve a high level of cooperation and service.
5. Anticipates company needs, market conditions, and materials improvements to insure efficient fulfillment of materials requirements.
6. Keeps buyers' guides up-to-date and accurate to achieve efficient requisitioning at the plant level.
7. Develops and institutes more efficient purchasing methods in order to promote economical and timely purchasing.

It is readily apparent from a study of these forms that there is a definite technique to writing a workable job specification. Make sure the fundamentals are there first, and then add the trimmings.

Buyer Training

It is virtually impossible to acquire trained and talented packaging buyers; no formal training areas for the position exist. The aspiring packaging buyer has a good start with a practical or scholastic background in packaging, a business administration major, purchasing training or experience, or, in some cases, sales experience. Since packaging buyers must be trained, the quality of the raw material is most important. The normal desire would be to promote to buyer from within and bring in an assistant buyer for training.

But what are potential areas for recruiting personnel? College graduates from one of the few schools offering courses in areas related to packaging are usually good gambles, and checking packaging R&D and design for potential candidates is a good idea. Salesmen of a certain temperament are often excellent prospects, and of course check other companies for candidates. Here again, the care devoted to the job specification allows leeway in hiring qualified prospects.

The training pace for an assistant buyer depends on the buyer's dedication to this task. Motivating the assistant to learn quickly and efficiently is a function of management. The major devices for training him generally consist of the following:

- Personal counseling. This should be handled by the buyer and the manager. Usually, the buyer handles technical matters and the manager imparts the concept of the position. Most important in this counseling is to convey immediately the type of conduct expected from all personnel. Professionalism in all actions should be the key word. Stress to the new assistant buyer that he is always a representative of the corporation, and warn him of temptations and pitfalls.
- Plant visits. If properly planned and handled, the visit to a source's plant is a most valuable tool. It must be a working visit, however, to have any real, lasting, training value. The shirtsleeves, dirty hands, share-with-the-press-crew type of visit works wonders; there is honest appreciation on both sides of the difficulties of producing quality packaging, and all are better for it.
- Industry seminars and schools. These are excellent vehicles of instruction. Almost any kind of formal education in the packaging purchasing area has great benefit.
- Company training classes. These should be a mecca for all personnel.
- Reading material. Matter of this nature, whether technical or concerning materials, equipment, purchasing management, marketing, or what have you, is important. Both the manager and buyer must constantly stress the never ending need to learn by stimulating the assistant buyer's intellectual curiosity and stressing the availability of the materials.
- Intracompany orientation. If the assistant buyer knows the organization, the available staff sections, and how to use them, it will make matters simpler for all concerned.

Buyer Function

Just how does the buyer work within the framework of the corporation? It is safe to say that a good buyer operates as efficiently as the system permits. He performs many functions, each related to certain departments or areas of packaging. The following is a brief checklist of functions as they interrelate with other departments. The purpose is to provide an outline of the dimensions of a multiplant packaging procurement function.

Buying. This is a basic function of the department and the primary reason for its existence.

Responsibility	*Major Department Involved*
Final artwork	Packaging design
Specifications	Packaging R&D
Estimate	Production planning
Position check (No checks are needed if the item is being produced for the first time.)	Plant and source inventories
Purchase order	Purchasing
Buyers' guides and clerical	Purchasing
Paying invoices	Accounting
Timing and schedules	Marketing, production, and sometimes advertising

Planning. This is one of the most important parts of corporate life. Plans used by purchasing are many and varied and require assistance from diverse areas.

- Cost-reduction plan. Packaging R&D, marketing, and engineering could easily be involved in this type of plan (possibly if a drastic change in materials were proposed); production's advice would be asked if machine speeds were to be affected and packaging design if the program involved reduction of colors.
- Material production plan. The buyer generally constructs this plan, hopefully after he has reviewed the promotions of his marketing group and established their effect on the

packaging area. If there are to be many changes, the buyer should work out a master schedule with marketing to establish key dates for the accomplishment of original designs, sketches, comprehensives, final art, pack start date (refers to the date that all packaging materials must be available), end date, and shipping and in-market distribution dates. Production planning plays a key role here because purchasing depends on the figures it provides for the actual production of packaging materials.

- Packaging R&D. R&D is involved in purchasing's planning because it must have the required specifications ready by a scheduled date.
- Long-range planning (packaging changes). This activity consumes much of the buyer's time if the relationship with marketing is as it should be. The buyer and a merchandise group should review the group's master plan for the year, and the buyer should specify the deadlines for the previously mentioned operations. The merchandising group must be alerted to the need to meet these dates by making prompt decisions.

Inventory control. This vital activity involves a lot of work with the personnel who control plant inventory, the sources that carry inventory for the corporation, marketing groups, and production people.

Interviewing sources. Meeting with sources to decide which ones to develop is primarily a buyer's function, but it often involves adding marketing, R&D, and packaging design or engineering personnel to the meeting.

These are but some of the major areas of the buyer's job. Added is the mass of necessary detail, checking, tracing, getting credits, handling losses, writing reports, and many other items. But one thing stands out in all these activities: In the corporate society today almost everything the buyer does involves someone else in the activity; hence the need for him to understand, direct, and motivate others.

Chapter VII

Operations Control in Packaging Procurement

CENTRALIZATION and decentralization are the only two pure conceptual approaches to packaging procurement. The decision on which to choose is a primary one that must be carefully made after weighing such matters as staff sequence, effective supervision and controls, R&D, design, field autonomy, costs as related to volume needs, technical expertise, extent of home-office rule desired, and a host of other factors peculiar to each particular business.

OPERATIONAL CONCEPT

The growth of corporations by acquisition and merger has given the packaging procurement function an importance distinct from that of equipment or from maintenance, repair, and operational (MRO) purchasing, which have, by the nature of their standard type of items, an ingrown flexibility and adaptability to either centralized or decentralized purchasing. Moreover, there are buying techniques that can make equipment and MRO purchasing quite efficient for the corporation. This also applies to most types of raw materials used in the manufacturing process.

In handling raw materials and MRO items, large multiunit corporations have adopted a compromise approach. The bulk of special plant needs are handled by an on-site purchasing group with buying authority. A corporate purchasing department assists the field or plant personnel and establishes and administers national contracts for common items. This approach lends itself to the utilization of systems contracting and central buying but still leaves room for local initiative. From an organizational standpoint and from the aspect of personnel growth and development, this system has great merit. The key to success is clear identification of each job's limits in terms of what can be bought on local initiative and what must be ordered against national contracts. The local units, while enjoying a great deal of freedom, are still under the control of the corporate office.

Local purchasing's reporting channels. Several alternatives for establishing the reporting status of the local purchasing group are possible. The first is to have the group report directly to the home office and receive all orders and instructions from there. But this creates immediate problems by removing control of an important profit-making unit from the local manager, by allowing headquarters to inject itself into the local management situation, and by making impossible an accurate estimate of the efficiency of local management because of the influence of corporate headquarters. Therefore, this technique should normally be avoided.

An alternative is to have the unit report to the local manager and be under his control. The home office has only advisory powers and can be overruled by local management, which is why strong local managers favor this approach. A certain amount of red tape may be avoided this way, but there is the danger that home office advice and instructions may be disregarded, and in extreme cases, harm can be done to the corporate purchasing picture by failure to support national contracts. The subject of national contracts arouses local managements, which are convinced that they can purchase better locally, but these managements tend to make decisions on local cost alone and not to consider such factors as corporatewide standardization, servicing, replacement, and overall pricing. Therefore, local control has grave effects on the company's utilization of national contracts and can also have considerable bearing on its personnel development programs.

A third method is a combination; the local purchasing group is part of the local manager's staff, and the local manager has a voice in buying items peculiar to his plant. Although the manager has administrative control, the home office has functional guidance over the group. But each local group's sphere of control must be clearly spelled out. It must never be put in the position of having to choose between masters. If communications are maintained between the local manager and the home office and concerned parties are kept abreast of events, then this system works.

Although the buying of MRO items does not affect packaging purchases, the MRO purchasing technique reveals how *not* to handle creative packaging procurement. The word "creative" means that the procurement of routine items, such as corrugated containers, can be left to local option if the purchase is justified. Organizational purists generally do not agree that a local buying group should have initial buying authority in a strictly centralized buying setup.

In general, the most efficient approach to corporate packaging procurement is strict adherence to the principle of complete purchasing centralization. Although this limits the options of the local unit, a little thought and effort permit an approach that accomplishes the necessary goals. Decentralization of purchasing, including packaging procurement, brings on problems in quite a few areas.

Decentralized Packaging Procurement

Decentralization, in its pure form, assumes that all the packaging needs for a given plant are to be fulfilled by suppliers selected by the plant's purchasing personnel. But the question arises: Can the local plant design and engineer the package by itself, or is central control necessary?

Buying power and costs. If only one plant uses a package and the volume is sufficient, local buying could work; the problem arises when several plants use a common package. Local buying reduces volume, increases the base price and the cost of plates and make-ready because of duplication, and contributes greatly to non-standard packages. No two printers work alike, even from the same art. Packages produced in separate plants are different, particularly in color match.

Staff services. Packaging R&D and design are staff services. If they are not handled on a centralized basis, what often results is:

- Duplication of staffs, resulting in increased personnel costs and generally a dilution of talent owing to budget restrictions.
- Potential legal problems arising from package claims not cleared through the legal department. The more power at the local level, the less the tendency to check with the home office.
- Rising package costs fostered by duplication of art, design, and development.
- Lack of design control, including the possible loss of corporate identity, which in itself is enough to negate the decentralized approach.
- Greatly hindered coordination caused by problems of communication and authority.
- Loss of continuity or control in marketing, which could be fatal. The communications problem is unbelievably difficult.

Duplication of staffs and efforts. Decentralization means duplicate groups in each location, and staffs working under the decentralized plan do not have the caliber of talent available in a centralized setup. Talent costs money—the better the quality the more money it costs—and local units cannot afford top talent. The problem of duplication of effort is probably the main disadvantage of decentralization. But the difficulties of standardizing materials and of headquarters' control over such items as budgets, inventories, write-offs, and losses also present formidable obstacles. Corporations whose plants have common equipment, designs, and needs should handle their purchasing requirements as a unit.

The general conclusion for packaging procurement, therefore, is that the type of decentralization where the local plant is authorized to handle all or even part of the package development cycle is not practical. The foregoing reasons are only a few, but they are significant, and, when they are pondered and analyzed, they build an overwhelming case against the theory of decentralization of packaging procurement.

Centralized Packaging Procurement

Centralization has all the benefits that are problems for decentralization. For example, communications are simplified because packaging design and R&D route the results of their individual efforts to purchasing, which becomes the field contact.

When a centralized system is being set up, purchasing should follow some basic ground rules. The classic illustration, which is a common one today, is where the parent company is acquiring diversified companies. This places the corporate purchasing group in a difficult position because each new plant is usually equipped with a purchasing department of unknown quality. The process of converting from an individual plant's packaging purchasing setup to a corporate centralized arrangement should follow a definite step-by-step procedure. The following approach is not the only way to handle the situation, but it is general enough to have universal applications.

The Integration Process

Establish the policy. First, and most important, top management must establish the policy for centralized packaging procurement. The decision to integrate plants and departments must be firmly made and transmitted to subordinate units and a rough timetable for accomplishment should be constructed.

Analysis of materials. A detailed plant study must be made that shows the entire picture of current materials used and the specifications for them; ordering frequency; source alignments; current commitments to be honored; invoice-handling techniques; graphics involved; plates or other equipment owned by the plant finished-packaging inventory, both in-plant and at sources; and stocks of raw materials, such as metal (tinplate), carton board, plastic roll-stock, and the like. Also, the supply of raw or finished materials, whether on order or in stock, should be evaluated to determine how long it will last.

Integration of materials. Because standardization of materials is a basic concept of volume purchasing, a very complete analysis and comparison of plant materials and specifications must be made to determine those that are similar or potentially similar—a process that may require some testing and analysis by R&D. Wherever possible, common or standard specifications for future purchasing must be set. Until this is done, it is quite impossible to complete a picture of the corporation's material requirements.

Organization and personnel review. When the range of involved materials has been determined, dollar-volume estimates completed, and corporate marketing policy regarding the new plant's products established, another study should be made on how best to organize for purchasing and where to use available personnel.

The records of each man involved in the plant's packaging procurement should be carefully reviewed and any interviews or corporate personnel tests completed. At this time, a decision must be made as to what to do with these people. The corporate staff might need enlarging, and, if possible, this should be done by transferring the plant personnel. As a rule, senior buyers are the most likely candidates for corporate responsibilities; younger men with potential should be left at the plants for training and can operate that end of the purchasing system. Personnel who do not seem to qualify should be separated from purchasing, but every effort should be made to reassign them to another segment of the corporation.

Establish the system. Everyone involved, both in the field and at the home office, must be fully informed, in writing, as to how things will operate. Policy sheets should be issued on specific matters. For example, employees should be told which reports will be required and how often. The written statement should include sample reports as guides. Since invoices may be paid locally or centrally, the policy on handling them must be clearly spelled out. If they are to be handled centrally, then the plant must be told how to handle receipts, which must be forwarded to the home office to be matched with the invoice. It is preferable to pay invoices locally, if possible, because this saves considerable time and paper handling at headquarters. But the system, of course, will vary with the corporate philosophy.

The policy must also be set for local authority. Normally, the plant package-control personnel should be allowed to place orders for materials against master contracts established by central purchasing. But any latitude permitted the plants beyond the contracts (the benefits of this idea are doubtful) should be detailed and expenditure limits established. Some jobs have minor local problems, such as imprinting information on a package, but these are usually emergencies caused by the need to correct errors on the package, and the plants are requested to handle the situation by central purchasing.

A vital factor in the centralized purchasing operation is the importance of having some standard means of getting the ordering data to the field. A "buying information sheet," which provides all the data necessary for the plant personnel, is a good way to do this. It can be formulated in numerous ways, but the various items of information should be numbered and an explanation of what each numbered item means should be printed on the back of the form for reference by field personnel. Following is a suggested setup grouping related items together. It can be altered to fit the individual company's circumstances.

1. Plants concerned. The plants to which the bulletin applies are indicated by an *X*. If all plants are affected the

"all plants" space is marked with an *X*, and no individual plants are indicated.

2. Function of the bulletin being issued. It contains the following designations:
   - Date. Date of issuance.
   - Serial number. Number assigned to this bulletin by the buyer. All bulletins are numbered.
   - Buyer. Name of buyer issuing the bulletin.
   - Replaces, revises, or new item. Applicable term is marked with an *X*. If this bulletin is either a replacement or a revision, then the serial number and issue date of the bulletin being replaced or revised are given.
   - Reason for bulletin. Change in either cost, terms, pack (refers to how the packaging material is packed for shipment), source, or other reason. Appropriate explanation is marked with an *X*. If "other reason" is checked, then a brief explanation is inserted in space allowed.
3. Vendor name and address. This is where the plant personnel send release orders. A phone number and name of a person to contact should be included.
4. Shipping point. Point from which goods are shipped.
5. F.O.B. point. Self-explanatory.
6. Terms. Applicable trade terms.
7. Transportation allowance. Any special arrangements on allowances are noted here.
8. Transit time. Average time needed for shipment to travel from shipping point to destination.
9. Package material number. This number, in conjunction with the item description (item 11), is used when ordering the item. It is for computer identification, and it could be as long as nine digits, depending on the EDP system.
10. Product code number. Code currently assigned to product line, if used.
11. Package material description. Specific description of the material involved.

12. Vendor price. Price allowed under contract placed by central purchasing.
13. Average freight per unit. Average cost of shipping one unit (item 16) from shipping point to destination.
14. Delivered price. Same as "landed price," which includes base price plus average freight costs.
15. Pricing unit. Refers to way that items are priced, based on contract with vendor. For example, price could be quoted in terms of thousands, dozens, pounds, bales, each, or in any other way peculiar to an industry.
16. Vendor pack.* This category breaks down into
    a. Number of units. The basic pack as put up by the vendor. The listing might read, for example: plastic bags, 4,000 per carton, weighing 23 pounds; or multiwall bags, 100 bags per bundle, weighing 40 pounds.
    b. Packing method. This contains numbers one through ten, each item relating to a packing method.
        1. Corrugated box
        2. Bundle
        3. Bale
        4. Nested
        5. Can
        6. Folded flat
        7. K D flat
        8. Roll
        9. Spool
        10. Skid or pallet
    c. Weight. This is related to the weight of the basic vendor pack. For instance, assume that the merchandise is plastic bags. If the information in 16*a, b,* and *c* read: 6,000, 1, 42, it would mean that the basic pack was 6,000 bags per corrugated carton for a weight of 42 pounds.
17. Special instructions. In this category would be listed anything special, such as skid loads or other pertinent data. If we use the preceding bag example in item 16, and if the cartons are loaded 12 cartons to a skid for shipment, the notation in item 17 would read "skid, 12." From this data one can compute the weight of the skid by multi-

*The numbers used in 16*a* must be related to 16*b* for a full explanation.

plying the weight listed in 16*c* by the number of cartons per skid in 17.

There are other written forms in addition to the buying information sheet. One of these is a bulletin reflecting the corporate policy on inventory reports and controls. It covers such items as frequency of reports, times per year packaging materials must be physically inventoried, insight and coverage recommendations, pricing of inventories, use of cullage factors in computing packaging costs at the plant level, and any other important inventory control factors. The bulletin should be as complete as possible and yet simple enough to be useful as a working tool.

If packaging materials are programmed into data processing equipment, a bulletin should be issued explaining in detail the numbering system, reports and components of reports that are to be generated by the computer, the kind of data available to the plants, and any other information applicable to central purchasing's concept of operation.

A workable scrap-disposal system should be developed and a statement issued on it. But the data required from the plants must be kept to a minimum. A five-part form, which requires only the following information, is recommended.

- Plant name and number.
- Item number (usually the package material number, which is often the data processing code number).
- Item description.
- Quantity to be scrapped.
- Explanation for scrapping; for example, over-ordered or production stopped.

EDP Applications

The computer offers a great opportunity to aid and assist such functions as forecasting, inventory control, and vendor performance appraisal. But the major computer applications to the purchase of packaging materials today generally revolve around inventory control. The plant and source inventories are brought together, and a corporate package materials inventory report is created. In most companies as the management information system progresses, the output from the computer increases. When a basic system for the use of the available data in the corporation is being established, guidelines must be set for processing and reporting frequencies. The normal categories for issuing data are: on demand, daily, weekly, monthly, quarterly, and yearly.

The major problem is how to plan for using a computer. It is difficult to analyze the daily operations and to ask the proper questions. So the following is a suggested, simplified approach to computerization of the package purchasing operation. It consists basically of two parts: the input required for each type of report; and the output expected from each type of input. This technique is somewhat simplified, but in general it is adequate for the average packaging procurement operation.

INPUT

On demand. The operations of package materials purchasing require that information, which is held on magnetic files, be accessible upon request. These data, mainly of a "status" and "position" nature, are needed to facilitate certain preprocurement decisions and to increase the effectiveness of inventory coverage policies. An on-demand subsystem should be designed to assist in fulfilling these requirements, and it should contain comprehensive files including: packaging materials inventory, forecast, and history; and both background and up-to-date information on buyer performance, accounts payable, purchase orders, and purchase contracts and commitments.

Daily procurement plan. Use of a daily procurement plan for package materials purchasing requires the inclusion of selected status, position, performance, and analysis data about given packaging materials. In order to secure the most economical price and the most favorable quality and delivery, the plan should have a daily procurement subsystem containing the same historical and up-to-date information as the on-demand system plus data on vendor performance.

Weekly status and analysis. A weekly assessment and evaluation of package materials procurement allows sufficient time to view the procurement cycle, which involves the purchasing decision, its execution, the effect of the decision, and reaction to the effect, and yet ample time to initiate corrective action if needed. Its subsystem should contain data on the package materials inventory, forecast, purchase orders, and contracts and commitments.

Monthly status, analysis, performance, and forecast. A monthly review covers a variety of items needed to evaluate performance, re-establish controls, and utilize forecasts. Current data must be available to accomplish a status revision and analysis. The monthly subsystem should have complete information on the same topics as the daily subsystem.

Quarterly status, analysis, and performance. Management is supplied with operational information by the quarterly status, analysis, and performance review. Its subsystem should provide the inventory and history of packaging materials and data on the history of buyer and vendor performance and on purchase orders.

Yearly analysis, performance, and forecast. Package materials purchasing requires yearly input relating to costs, volume, usage, and performance in order to evaluate its history and plan its anticipated long-range position. The yearly subsystem should cover the package materials inventory, forecast, and history, background on buyer and vendor performance, and purchasing contracts and commitments.

Processing

Operations of package materials purchasing are designed to utilize the most current procurement information, which must be available from computer magnetic files maintained and updated daily. Although all packaging materials are not purchased daily, there is always some buying. Therefore, in order to implement a daily procurement plan on any given package material or category, the system gathers and updates source information once a day. The origins of the data are mainly within the company; however, some outside sources are used. Among the more significant of these data acquisition sources are package materials purchasing, production, traffic, product management, plant packaging materials coordinators, and, from the outside, vendors and trade publications.

Output

On demand. Output data available on a demand basis are designated for distribution to purchasing. Selected data go to specific plant package materials coordinators. The following are samples of the kind of output that can be expected:

- Inventory status, including a breakdown of how much each plant holds of one item and the total.
- Package materials corporate inventory.
- Anticipated inventory—position and coverage.
- Package materials savings report.
- Purchasing history analysis; frequency of purchases.
- In-transit order status.
- Purchase orders—status and position.
- Contract and commitment position by vendor.
- Effects of price increases or decreases.

Daily procurement plan. Daily output distribution is also aimed at purchasing, with selected data going to specific plant package materials coordinators. The data cover:

- Approved orders and requisitions.
- Plant inventory status.
- Notice of reorder situation.
- Expediting data (phone numbers, routing lists, and the like).
- Notice of late shipments.
- Performance of selected vendors.
- Performance of selected buyers.
- Production plan changes by packaging item.

- Coverage of following week's production plan.
- Shortage caused by changes.
- Package materials buying guide.

In addition, the daily output updates files on the package materials inventory, forecast, and history, as well as on buyer and vendor performance, accounts payable, purchase orders, and contracts and commitments.

Weekly status and analysis. Output distributed weekly is composed of status and analysis material that is routed in the same way as the on-demand and daily data. This information is designed to show:

- Actual dollar contracts and commitments.
- Contract status (coverage).
- Schedule of package materials changes.
- Deal materials status.
- Package materials inventory status and usage.
- Revised bill of materials.
- Analysis of in-transit order status.

The files updated include those on purchase orders, contracts and commitments, materials forecasts, and inventory.

Monthly status, analysis, performance, and forecast. This review's distribution is again handled in the same way as the preceding statements. Its goal is to reveal:

- Corporate plan versus actual expenditure.
- Production plan (estimate).
- Package materials purchasing forecast.
- Corporate inventory (by buyer).
- Plant inventory position and status.
- Summary of local plant purchases.
- Obsolete inventory.
- Scrapped packaging material.
- Materials catalog.
- Performance-trend charts.

The package materials inventory, forecast, and history are made more current.

Quarterly status, analysis, and performance. This output usually goes to central purchasing, but not, as a rule, to the plants, because it does not have a strong bearing on their operations. The data deal with:

- Analysis of open-order status.
- Package materials master information.
- Small quantity purchases.

Also, the files are updated on the history of buyer and vendor performances and purchase orders.

Yearly analysis, performance, and forecast. The output from the yearly review is distributed in the same fashion as the quarterly data. It deals with:

- Anticipated dollar commitment.
- Corporate plan and adjustment.
- Vendor evaluation.
- Dollar volume (by vendor).
- Buyer performance.
- Package materials inventory status forecast.
- Performance-trend charts.

The files made more current include the package materials inventory and forecast, the history of buyer and vendor performance, and purchase contracts and commitments.

This then is a plan that can handle the job of utilizing the computer in the package procurement operation. The computer is the way of the future in the purchasing business; it is the only answer to the avalanche of paper and detail that engulfs everyone.

Chapter VIII

Corporate Packaging Management

EVERYONE INVOLVED in packaging recognizes the massive range of problems encountered in the development of a given package. Yet few can clearly delineate just what causes the delays, the mistakes, the omissions, and the schedules that are twisted out of their original shapes. To be sure, there is no simple answer, no panacea that is the perfect system or solution to the management of the packaging function.

Much of the success of any given organizational approach depends greatly on the strength, knowledge, dedication, and viewpoints of the key members in the packaging operation. The relative strength of the major departments involved often determines the eventual path that the procedures take. When any kind of a package-handling system is being set up, it is always wise to build in some checks and balances. They tend both to maintain a certain power equality between units and to promote better work because an outsider, generally unsympathetic, is available to catch errors that might have been missed the first time around.

The major problem in organization seems to center around how the corporation chooses to attack the packaging task. As we have seen, there are several basic ways to arrange and locate the three key departments in the packaging effort. Each approach is operable, but some efforts are more successful than others. The advantages and disadvantages of each should be viewed with an eye to the overall corporate objectives before a major overhaul of the company's structure is attempted.

Three Independent Departments

This is perhaps the most widely used of the organizational techniques because it is easy to organize, but it is extremely difficult to control. Under the usual arrangement, the package procurement people reside in the purchasing department, R&D is usually attached to the laboratory, and design probably reports to either the president, the vice president of marketing, or the vice president of advertising. There is no rule pinpointing exactly where these units should report, but it makes little difference. Changing the reporting structures does not alter or alleviate the problems that are inherent in this approach. It can work well in certain corporate climates, but it fails completely in others. Let us look at some of its major difficulties.

People. Personality impact is greatest in this type of organizational setup. Each unit is intent on preserving its identity, on growing, on expanding its sphere of interest, and eventually on becoming the dominant figure in the triangle—a tailor-made situation for the emergence of a strong personality as leader. Unless there has been much soul searching and much understanding between leaders, the process dissolves in absolute chaos. The stage is set for rivalry and jealousy over operational areas. And there is always the tendency, under stress, to employ the normal checks and balances in this system as weapons.

Interest areas. In an organizational structure like this, it is inevitable that the areas of interest overlap and that incursions into sacred departmental confines occur. R&D is always in conflict with design over the coordinators, and for some strange reason design directors feel obligated to inject themselves into the physical development of the package. These forays into R&D's realm of influence are guaranteed to set off a violent exchange of memoranda accompanied by some bitter words that are never really forgotten.

Design runs afoul of purchasing in the same way. The coordinators, acting as development people, begin contacting sources, placing sample orders, and in some cases trying to adjust schedules at converting plants in order to secure manufacturing time for tests. Purchasing's general reaction is to chop off this extraneous activity by: (1) refusing to honor invoices that do not carry a purchasing department order number, (2) advising design personnel to clear all requests for source action through the buyer involved, and, most effectively, (3) ordering sources to respond to all direct contacts initiated by design people with the statement that their requests must go through purchasing. Sources recoil at this last technique until they realize that purchasing really controls their destiny.

Purchasing is somewhat at the mercy of design, which normally provides the final art. Schedules are rarely met, and design has an endless supply of excuses for delays. In this case purchasing is left with insufficient time to do the job and invariably takes the majority of the blame for the upset schedule—a fact that causes extreme resentment in purchasing and creates an atmosphere of general apprehension wherever design is concerned.

Purchasing and R&D have their confrontations over the quality of specifications and the day-to-day working arrangements with sources. As a rule, the final working arrangement has purchasing carefully select the source with the needs of R&D in mind, make the source available for R&D's use, and review and consult on all specifications before they are issued.

In spite of having the seeds of disaster built in, the independent department system can work effectively, but certain basic steps must be initially taken to avoid the potential conflicts. First, company officials, particularly those in personnel and those heading the key departments, must lay down the rules. They must circumscribe each group's areas of influence, the limits of its authority, and its responsibilities for supplying specific support efforts. The next step is to structure the departments' organizations with an eye to mission accomplishment and mutual support. Every possible action should be taken to clarify each function and its relationship to others, to spell out specific assignment areas, and to simplify the burden of communications. When this has been done, the job specification for each position should be written. It is the best guide that a man has as to what is expected of him in the task assigned, and it should reflect the scope and limitations of his work and authority. If this is not done properly, the stage is set for trouble. The last assignment is to work out an explicit, detailed procedure for step-by-step handling of packaging projects. The function of each department must be fitted into the packaging setup at the proper time sequence. Failure to establish and police the system will merely compound the confusion already inherent in packaging management.

Although these rules are vital to the effective functioning of a separate department approach, they are fundamentals for any system. The groundwork must be solid, utilizing the strengths of all areas, and must provide for more than adequate communications.

A Staff Function

Another school of thought feels that packaging can be better handled by a corporate packaging management group that controls the functions of packaging development. In essence, this philosophy views packaging management as a corporate staff function. Those who favor this technique say that the creation of a top level staff provides a place to solve many of the conflicts present in other systems. The staff acts as a collecting agency for reports on such matters as market changes, consumer reactions, competitive activities (particularly in graphics), new equipment availability, company changes in products, position checks on various products regarding their share of the market, and a host of other related items. Having access to all this data, the staff's major job is to assist product development and marketing people in determining packaging project priorities and, hopefully, to attend to some long-range package planning. Inherent in the staff philosophy is the hope that it would function as a neutral mediation agency to get packaging decisions speeded up so that development programs would not be unnecessarily delayed.

There is some difference of opinion as to how closely connected the staff should become with each of the major packaging development units. One idea is to use the staff simply for collecting marketing and competitive data essential for future planning, to set priority schedules for packaging projects, and to act as mediation agent and general clearinghouse. Another viewpoint encompasses this description but goes far beyond it to involve the staff deeply in the work of each major packaging department. Such depth of involvement would extend to control of the design, consumer tests of designs, trial production runs, physical properties of the package, recommendations and approvals of equipment, and, in general, to control of the corporate packaging effort. This deep-involvement approach faces instant disaster for a variety of reasons. The minute any type of staff man attempts to assume some measure of direction or control over a line function the reaction is usually violent. Moreover, such deep involvement entails practically a duplication of effort in the key areas of purchasing, R&D, and design. The results do not justify the expense, and the overall efficiency of the involved departments probably suffers. However, there is no doubt that, in certain management climates, the staff idea works, but such an approach has limited appeal.

Since we live in a world of compromises, and if a corporation does not want to jump completely to a line approach where the three major elements are controlled by one corporate vice president of packaging, there is an intermediate approach.

THE NONSTAFF PACKAGING TEAM

Many corporations that cannot yet face the prospect of packaging as a major line function favor a team concept. A packaging team consists of a nucleus of three permanent members, the managers of packaging purchasing, R&D, and design. Each of the permanent staff should be backed by an alternate so that each key department is always represented. Additional workers should be available from engineering, production, consumer research, merchandising, legal, quality control, and product development as required. The packaging team should set priorities and schedules, formulate information requests supplied by associated areas, make decisions on physical specifications, designs, and colors, prepare copy or supervise its preparation, review proofs, get all necessary clearances, and establish color standards. These are only the main functions of the group; they perform or have performed hundreds of necessary supportive actions. The team calls and controls all meetings on current projects and determines the homework associated areas are to complete before the next meeting. It should present a powerful front because it represents the bulk of packaging knowledge. And this has the desirable effect of forcing the three leaders to function as a mutually supportive unit.

The key to the team's success is to establish and advertise it throughout the company as a corporate working unit accountable to the chief executive. This bit of organizational psychology eliminates most of the bickering between the design people and the merchandisers, who fancy themselves art directors. Often, if there must be a meeting of minds over a design, this can best be accomplished by the packaging team, which can call on its superior group experience; moreover, in rare cases of absolute deadlock, it has access to the president as an arbiter. The knowledge of this alternative may well temper the fervor of the amateur art director/merchandiser, as may the need to contend with three adversaries instead of one who could more easily be browbeaten.

The team idea then has much merit. It is relatively new and

Exhibit 10

Organization of the Purchasing/R&D Department

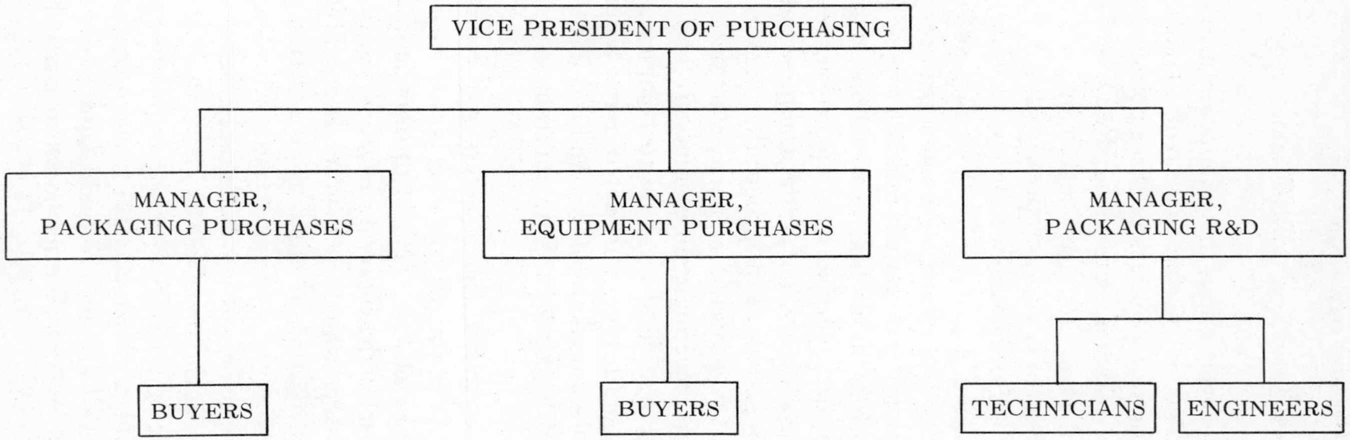

untried, but it will certainly develop in the future. The main secret of its success is proper location and reporting status—that is, it must report directly to the top. The three key members must work out the operating techniques. They must work as a unit, thrash out their differences in private, and always present a unified front to the rest of the corporation.

The Purchasing/R&D Team

In recent years, another plan of attack on the packaging problem has evolved—the union of purchasing with R&D. When the conflict between the two gets so bad that the work is not being properly done, then top management must step in and perform a little organizational surgery. The usual result is the incorporation of R&D into purchasing. The initial shock to R&D is the cause for some complaining and, eventually, some soul-searching, but the union makes a lot of sense because the combination of talents should produce a better job. Rivalry is eliminated because there is no need to struggle for domination, the benefits of single, overall leadership for planning and control cannot be denied, and purchasing's technical knowledge and industrial contracts can be better utilized by R&D. The department, which would be a subdepartment of purchasing, would have quite a simple setup (see Exhibit 10). The rationale for this type of merger varies, depending on whether you view it from purchasing's perspective or from R&D's side. Purchasing generally defends subordinating R&D with very strong arguments that are so close to being policy statements that the position is hard to attack. Basically, the reasoning is as follows:

- The first task of the corporation is to show a profit.
- The purchasing department, by exercising expedient control over expenditures, can make a very sizable contribution to the profit picture.
- The cost of packaging, which is a very real part of the cost of the merchandise, is a factor over which purchasing exercises control.

- One thing that causes purchasing to lose some control over the cost of packaging is the specification prepared by R&D. Therefore, purchasing should exercise some control over the specification and other aspects of the package. Unrealistic specifications have been a major part of excessive packaging costs for many years.
- Source facilities can be better utilized with less friction if everyone is on the purchasing team, because purchasing controls the sources.

If, however, packaging procurement is a dominant part of purchasing's overall expenditures, R&D may well be placed in the package-purchasing section. Many packaging people feel that this approach offers the best opportunity for a unified operation and gives the best final results.

The future will probably see much more of this purchasing/R&D merger. It makes good sense organizationally, produces better results, eases control and planning, and opens up areas of growth for R&D. In addition, the benefits of cross-training, concentration on cost-saving projects, individual development, and overall improved performance cannot be overlooked. One of the major personnel problems of R&D is its potential for "locking in" employees. Many trained R&D men are excellent possibilities for becoming buyers of packaging materials, but in a divided setup, getting out of R&D and into purchasing is often quite impossible.

The Consolidated Department

Traditionally, packaging management in the corporation has not really been recognized as a management function or management need. Decentralization of effort has been the key note, with the three major packaging units operating as staff, or support, rather than as line units. The importance of packing in marketing and sales strategy, the increasing complexity of the packaging industry, the growth of materials and equipment available, and the mushrooming size of individual corporations have created the new philosophy of treating packaging development as a management function with a line concept. And this realization, that packaging in all its aspects is as recognizable a function as production, marketing, sales, legal, or engineering, has fostered the growth of the consolidated department.

The concept is simple—packaging management is an integral function of the corporate structure, and it deals with a highly complex and difficult-to-manage area of the business. The basic elements of the organization chart remain the same, but the functions are consolidated and report to a single head (see Exhibit 11). The important changes here are that the packaging function has an identity within the corporation, its leader is assigned vice presidential status, and the solidity of the arrangement is further insured by having him report to the president. This reporting route is the real key to success; if the vice president of packaging does not report directly to the top, the plan will fail. He will often be called on to arbitrate disputes between merchandising and packaging and must be as objective as possible. He cannot do this if he reports to merchandising. And if he is to report to merchandising why give him vice presidential status at all?

The next problem is how to structure the unit to do the best job. A reasonable start is outlined below. Of course, many variations are possible within the four major units, particularly within purchasing and design.

Purchasing. If purchasing is using the product buying team approach combined with some industry alignment, or even the straight product approach with no industry lines, then the coordinators should be assigned to purchasing. This enables the team leaders to organize better, to communicate more closely with the merchandising groups, and to train their people better. It also opens the door to the buying ranks for coordinators. Coordinators should be attached to product buying teams only; the industry lines operate with buyers and assistant buyers only. A typical setup within a product buying team would look like Exhibit 12. Under this type of structure the coordinator can really begin to function as he should. He becomes the liaison between merchandising, design, development, and purchasing. He should function

Exhibit 11

Organization of the Packaging Department

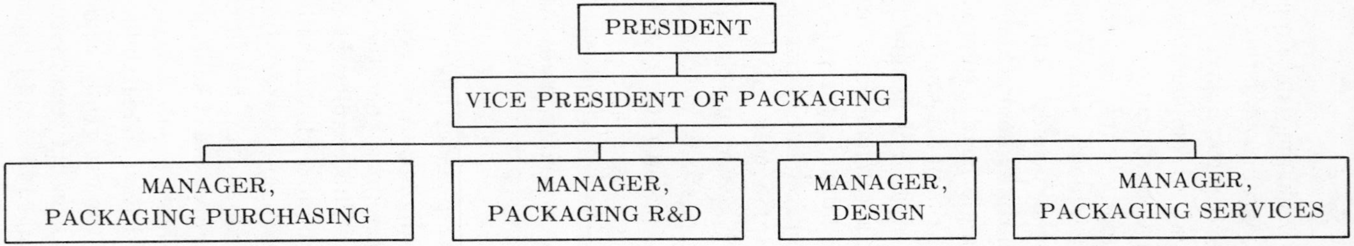

PRESIDENT

VICE PRESIDENT OF PACKAGING

MANAGER, PACKAGING PURCHASING

MANAGER, PACKAGING R&D

MANAGER, DESIGN

MANAGER, PACKAGING SERVICES

Exhibit 12

Organization of Buying Team--Ready-to-Eat Cereals (Merchandising Approach)

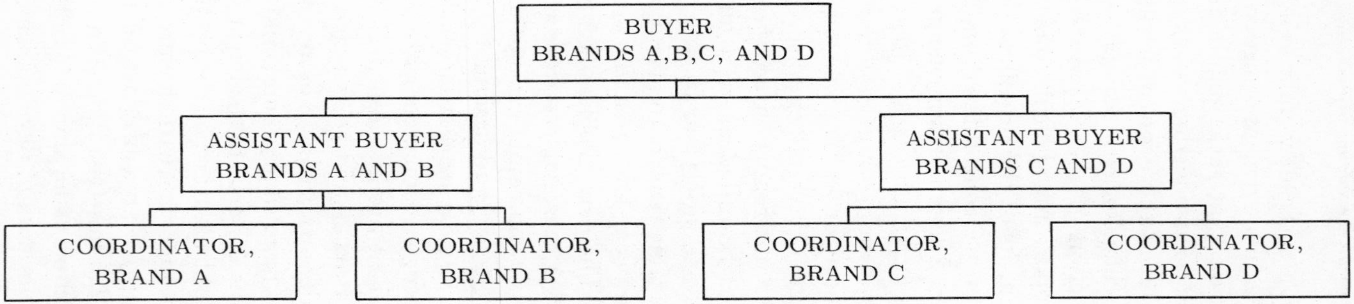

BUYER BRANDS A,B,C, AND D

ASSISTANT BUYER BRANDS A AND B

ASSISTANT BUYER BRANDS C AND D

COORDINATOR, BRAND A

COORDINATOR, BRAND B

COORDINATOR, BRAND C

COORDINATOR, BRAND D

as a planning assistant to the brand merchandiser, watching over his project schedules, getting art to him on time, watching the in-stock position in order to plan subsequent promotions, checking and hurrying development on projects, and absorbing all he can in the way of packaging knowledge. In short, he makes the wheels move quietly.

Packaging design. If the buying setup is by industry, then the coordinators are best left with packaging design. Design could operate with or without coordinators, but its sphere of interest depends on whether or not they are present. If they are not, then the department concentrates on creative design, acquiring or producing comprehensives, selecting color, and preparing final art. The major personnel involved are an art director, assistant art director, production chief, copy chief, and copywriters. If the coordinating task is added, more personnel are needed to do the liaison work.

Packaging services and schedule control. Packaging services is a rather new idea, an outgrowth of the purchasing services concept, and it is essential to a packaging department approach. Services handles matters that a department controller would; for example, the EDP setup for management information systems, budgets and reports, invoices, inventories, and a host of other service functions. The concept of setting aside a separate activity for schedule control is also new to most corporations. But with the myriad projects, there must be an easy way to keep track of each job's progress.

The consolidated department appears to be the best bet for the package-oriented corporation. The mere fact that the corporation realizes the difficulty of packaging management, grants packaging recognition as a distinct corporate function, and then provides the power and prestige to do the job immediately eliminates a host of problems. There are organizational problems, to be sure, but the concept is strong, organizationally right, and structured properly in the corporation. Such an approach succeeds.

Chapter IX

Packaging Procedures

UP TO THIS POINT, the various aspects of the jobs and departments involved in packaging have been highlighted, problem areas outlined, and various organizational approaches discussed. But a peculiar aspect of the packaging effort is that no matter how the corporate organization is arranged, the necessary and required functions do not vary; the only things that change are the techniques and relationships established to complete the job. It is essential, therefore, that the procedures required to produce packaging be reviewed in detail with the purpose of developing the packaging cycle and fitting the contributing parts into their proper places.

In this chapter the four-meetings plan will serve as the basis for directing the flow of packaging through the corporation. In most corporations it seems that the bane of the packaging effort and the major deterrent to programs are meetings, which are ostensibly held to discuss problems and make decisions. In general, they fall far short of fulfilling their purpose. Packaging meetings fail for many reasons, but the principal causes are lack of information and, particularly, lack of properly done homework.

Packaging decisions are no more difficult than other decisions, but they do not get made on time simply because people do not like to make decisions. This fact must be recognized, and the packaging procedure must be set up in such a way as to force decisions to be made. This is the basic premise behind the four-meetings plan. An individual is assigned at one meeting to see that certain decisions are made by the next meeting. The theory is simple: Decisions are made if pressure is applied, and the best kind of pressure is group pressure. Each individual is put on notice that failure to deliver will be resented by the group. This technique is not new or original, and it is used almost daily elsewhere in the corporation. The four-meetings plan may not be workable for every occasion, but if it serves to spotlight the key decisions that must be made on schedule, and gets those decisions made, then it may be considered a success. It sidesteps the problem of making one individual a gadfly to those who must make up their minds, because the group, which is a far more powerful and formidable force than the individual and therefore more difficult to buck, is the overall authority.

Package-Flow Chart

Before launching into a discussion of the four-meetings plan, it might be helpful to capsule primary tasks essential to producing a package. The sequence of events falls under a variety of names, but in reality it is no more than the package-flow chart. Once the required actions have been set up in relationship to each other, they can be fitted into the four-meetings plan.

Approval of the marketing objective. This merely means that the merchandising director obtains approval to proceed with a project.

Preparation of a statement of packaging objectives. No new project involving design and physical packaging development can get under way with any hope of success without a well-defined statement of objectives, which is, in effect, a brief review of the product involved (if new) and a careful outline of design objectives. The market to be aimed at is important, and the impression to be created is vital for design concept. The copy platform should be clearly outlined, covering such areas as age group to be reached, special characters to be created, and any tie-in planned with existing products. Of major interest are such things as special brand personality, new trademarks, and comparisons with or contrasts to competing products. Most important is the type or direction of appeal to be made to the consumer. The relationship of a new product to current ones should also be delineated in detail.

The final information needed, not so much for design guidance as for overall project evaluation and priority, is a financial review. This outline relates such factors as potential sales by units, desired share of market, estimated profit, and total sales contribution. These of course are projections, but they are of considerable importance to insure the interest of all involved parties. The document cannot be overemphasized, and yet it often represents one of the real weak spots in the system. The more complete the statement of objectives, the easier the tasks of design and development.

Review of the packaging objectives. This is the vital meeting that starts the actual package on its way. Present should be the merchandisers involved, the manager of design, the manager of R&D (if a physical shape must be developed), the manager of packaging purchases, and the designers. This is the time to ask questions, establish any necessary cost limitations on the finished package, and set the time sequence for the project—the initial pack date, distribution time, and market introduction date should be the guidelines. The amount of consumer-reaction testing, either on product or packaging suggestions, should also be determined at this time.

Design cost and time estimates. Before any design work is contracted, the manager of design or the art director should get an estimate of what the design work will cost. These costs should be carefully spelled out for each phase of the work, covering sketches, package mock-ups, comprehensives, and final art. If the studio is to write the copy, then this cost should also be included, and the time required for each step should be given. The design schedule is a guess at best, and it can be severely damaged by failure to capture the desired concept, by delays in decisions, or the like. If the overall schedule is going to falter, then this is where it will start. But a good statement of objectives that leaves no doubt as to what design route to pursue will probably prevent delays.

Approval of design cost and time. Merchandising must approve the estimate because the design costs are normally charged to its budget. How these charges are handled varies in almost every company, and it is not really a matter needing standardization, but since merchandising pays for the design work, it should be allowed to approve or disapprove the cost estimates.

Physical packaging study (if needed). At this point, R&D starts developing the physical shape and specifications from which purchasing will prepare its cost estimates. As a rule, the first specifications are for costing only, and they usually are adjusted before issuing final specifications.

Design program. The design manager, assuming approval of the cost estimate, assigns the job to a designer. The normal procedure is to set a schedule for presentation of initial ideas in sketch form. From these a more finalized design will hopefully be developed.

Review of initial design sketches. This review indicates whether the designer has really grasped the concept from the objectives. It is essential that these designs be shown to merchandising and purchasing, which is usually the case, but if this is not being done, it should be. The alternative is to let the design manager do the evaluating alone. Logic seems to dictate that merchandising, which must sell the product, and purchasing, which must reproduce the design, should be consulted. When agreement has been reached on the design elements, the artist can proceed to a more final version for review at the next session.

Receipt of tentative physical specifications. Purchasing's first real look at the dollar-and-cents aspects of the package comes when it receives the specifications from R&D. If, after review, purchasing approves, it will cost out the specifications. If there is any disagreement, then this is the time for the two departments to iron out problems—not when final specifications have been prepared.

Tentative costs. Because some vital components are missing at this time, the first costing effort is usually of a tentative nature. Purchasing, in all likelihood, does not have accurate production estimates and does not know the final number of colors or the complexity of the design. These are often referred to as preliminary, or class *B*, costs.

Receipt of final specifications. The time lag between preliminary and final specifications can vary greatly. If the preliminaries are in order, they can simply be restamped "final" and distributed.

Design finalization. The design manager should have the final comprehensive ready for review and approval by all interested parties. Accompanying the comprehensive should be the final copy sheets, actual color swatches, final packaging specifications, final engineering specifications (if any new equipment is required), a template of the actual package size, and a layout for the final artwork. It is wise to have the comprehensive reviewed by purchasing prior to submission to the merchandising group because there is always a difference of opinion between design and purchasing over the colors that are reproducible. Purchasing's pre-evaluation of the package can save a great deal of later embarrassment and can insure that the package will come out looking the way everyone wants it to. Often, purchasing does not get to view the comprehensive but sees only the final art and then, at this late date, must advise the art director that the package cannot be produced as designed. But at this point, everyone has approved it except purchasing. This situation occurs constantly and further complicates an already difficult business. Advance planning, interdepartmental cooperation, and better communication can avoid the obvious pitfalls.

Final design approval. The merchandising people's prerogative is to view the final design and issue the final approval. Actually, this should not be a problem if the designer followed the concepts approved in the preliminary stage. Approval should be automatic, but here again problems arise because people change their minds betweeen approval of preliminary design and receipt of final design. This presents the design manager with a difficult problem, and in the ensuing argument he should be supported by purchasing. Once approval is obtained, the preparation of final artwork begins.

Final costing by purchasing. When the final design is approved, the colors and layout determined, type specifications established, and final production figures set, purchasing should do the final costing as a basis for the last appraisal. Although it may seem too late in the game for costs to create problems, it can happen. Between preliminary and final design stages many changes could have occurred that might alter the cost of the package. Colors, for example, can be most expensive if they are not properly handled.

Needed project approvals. Every company has some kind of final approval system. In general merchandising houses this system is relatively simple, consisting of the buyer of the particular merchandise being packaged and the department head. In food companies the procedure tends to be more complicated because more top executives get involved. This is understandable, to a point, because the package has much more importance in the marketing and sales plans of a food corporation. The extended approval system does not generally cause too many problems, excepting delays in the schedule, as long as each signing executive does not appoint himself as an art critic. While a design is being routed for approval, all executives should realize that the final design is a combination of the packaging talents of design, R&D, purchasing, and merchandising. It is too late at this stage to start criticizing design or color.

Completion of final artwork. Purchasing uses the final artwork or mechanical to make the plates needed to print the package, so the final art should be reviewed and approved by design, purchasing, engineering, and merchandising. Other approvals might also be required, such as from the legal department.

Location of the source and purchase of the job. Purchasing, working with the final artwork and specifications, buys the job according to the system it uses. From this point on, it has complete control of the package.

First proof. This proof, which is the first color rendition of the artwork, is reviewed by purchasing, design, and merchandising. Corrections are made and the proof is returned to the source for either production or reproofing. It is wise to have the platemaker at the meeting so that he can get the comments firsthand, particularly if some plate rework is necessary.

A word of caution should be injected here as to the limitations on corrections. Usually, the corrections are confined to color tones and quality of reproduction. But too often, questions arise as to whether the photograph is really what is wanted. The platemaker and source have an obligation to match the artwork as closely as possible. If this is done, purchasing should entertain no radical changes. When the artwork reaches purchasing it is supposed to be correct, approved, and ready for production, so purchasing should not be responsible for the quality of the photograph at this stage. If, however, design or merchandising do not like the shot for reasons other than failure of the source to match the art, then purchasing must insist on a new photo that will produce the desired effect. This is necessary because the final art is the only standard for future comparison. A common problem is that at the proofing stage everyone decides that purchasing should attempt to alter the reproduction by making radical press adjustments when what is really needed is a new photo.

Approval of final proof. Once the final proof has been accepted, it becomes the guide for the printer as far as color and quality of reproduction are concerned. It does not, however, become the color standard. As mentioned earlier, proofing colors are different from those obtained under actual production.

Preparation of color standards. If the package is new, color standards should be pulled for the front and back panels. If, on subsequent runs, only back panels are changed, new standards may or may not be pulled.

Distribution of standards. Approved standards should be sent to all plants so that quality control can check the incoming goods. It is a good idea to have each plant view the standards in the same kind of lighting under which they were approved. Light changes color values, and unless conditions are the same there will be problems. Purchasing should have a light box, either commercial or specially built, for viewing proofs and standards. The same kind of unit should be installed in each plant and the quality control people trained to evaluate standards properly. Another suggestion, which generally arouses objections, is to have everyone associated with approving colors tested for color blindness. This includes buyers, art directors, pressmen, and quality control personnel. The number of cases of color-blind people involved in packaging is almost unbelievable. This recommendation might seem peculiar, but it is a good security measure that can avoid a great many future problems.

Production of the package materials. This is the actual production stage where the package is printed. It is wise to have a purchasing representative at press side to approve the result after the color is run up. This lessens the risk for the source and insures that the package will be as nearly perfect as possible.

The Four-Meetings Plan

Hopefully, the four-meetings plan will either provide an outline for taking the kinks out of a company's current packaging system or assist it in setting up a new approach to solving the problems of packaging management. The existence of the plan does not mean that smaller meetings will not be held to solve specific problems. The plan's objective is simply to get everyone who has to make a packaging decision or contribution to do his work on schedule.

MEETING ONE

Main Purposes

This is the initial planning meeting for the development of either one of the following:

- New package concept for a new product or group of products.
- Complete redesign of an active line of products.

The meeting will concern itself primarily with the following:

- Review of the statement of packaging objectives prepared by the merchandising group. (Of course, the marketing objective must have been approved and a statement of packaging objectives prepared *before* the meeting.)
- Gathering of basic data necessary for future work.
- Establishment of project schedules.

Those who should attend are the managers of design, R&D, packaging purchasing, and the involved merchandising department. There should also be representatives from production, consumer research, product development (if a new product is involved), and engineering.

At this meeting merchandising should give a rundown on its goals and aspirations for the project. The statement of objectives should be carefully discussed so that there are no doubts as to its meaning. Critical dates, such as initial pack date, trade distribution date, advertising-break dates, and introduction dates, should be firmly established. It is here that the schedule begins to take shape, but enough time must be allowed to complete the packaging cycle. The date for meeting two should be established.

Preparing for meeting two. Since the design manager is responsible for the coordinating staff, he sets up a framework for the project after the first meeting. Included in this skeleton is the establishment of a job reference number and a job file where all correspondence and copies of interdepartmental memos will be kept. And because he has overall responsibility for coordination, the design manager must receive copies of all related correspondence. The job number becomes vital, not only as a reference number, but as an internal security measure. The design manager also prepares the operating schedule that establishes periods for all phases. This is done by referring to the following master time chart, which indicates the average amount of time needed for the various stages, depending on the type of packaging involved.

MASTER TIME CHART

ITEM	*TYPE OF PROJECT*			
Category	*Complete New Artwork*	*Major Redesign*	*New Back Panel*	*Line Copy Change Only*
	(All given in weeks)			
Wrappers	8	8	4–5	4–5
Wrapper-rotogravure	12	10	10	3–6
Can label	8	8	8	4–5
Perforated BG can label	9–13	9–13	7–9	7–9
Carton	8	8	6–8	3
Films	6	6	4	4
Pouch paper	6	6	4	4
Small bags (process color)	8–9	8	6	6
Rollstock-rotogravure	9–12	9–12	—	4
Small bags (line printing)	6	6	—	5

The times indicated reflect average time lapses and may vary by corporation.

The design manager is also usually required, as part of the coordinating function, to submit to the R&D manager the request for a physical packaging study. This officially puts the job on the R&D schedule. The request can take almost any shape, but the following is a general format suitable for most job initiations. It is presented as a guideline; each company can tailor a form to its own needs. However, the request is a vital part of the packaging communication system.

REQUEST FOR PHYSICAL PACKAGING STUDY

TO: *Manager of Packaging Research and Development*
DATE:
ITEM:
OBJECTIVES:
ESTIMATED TEST-MARKET VOLUME:

TEST-MARKET DATE:
ESTIMATED NATIONAL VOLUME:
ESTIMATED DATE:
ACTION REQUESTED:

FIRST REPORT REQUESTED:
COMPLETION DATE:
SUBMITTED DATE:
APPROVED BY:
Manager of design
Manager of merchandising unit

The design manager must also get the bids on costs, have them approved, and finally assign the job with a date for completion of initial design sketches and preliminary copy.

The manager of packaging purchasing does not have much to do at this stage except assist the manager of R&D with sources. However, he should be prepared at meeting two to advise on rough costs and on printing or material procurement problems.

Engineering and production should conduct an investigation on what problems, if any, may be encountered. If the statement of objectives indicates a desire for a package not producible on current equipment, then this should be considered. However, until there is some indication of the package's physical shape, not much can be done.

Consumer research should be developing the final plan for any consumer testing that it wishes to do. By meeting two, it should have set its dates, decided on test-market areas, if any, and be ready to fit into the schedule.

Product development (if involved) should be putting the product in its final form, if this has not already been done. Sometimes packaging development is started before all the bugs are out of the product.

Upon receiving the request for a physical packaging study, R&D should begin work immediately. By the time meeting two arrives it should have several suggestions on suitable packaging types. The number of suggestions varies by product, industry, and corporation. Very often, particularly in the food industry, the tremendous investment in filling and packaging equipment precludes the development of ultra-exotic packaging. Conversely, in general merchandising houses, R&D usually has much more creative latitude. For the next meeting, then, the R&D manager should be ready with suggested prototypes of packages, preliminary material specifications, and, with the help of purchasing, early cost figures.

MEETING TWO

Main Purposes

- Review the initial design sketches and copy.
- Review preliminary physical package suggestions.
- Review production's and engineering's preliminary analyses.
- Review consumer research plan.
- Review product development progress (if applicable).
- Update the schedule.

Those who should attend are the managers of design, R&D, packaging purchasing, and the involved merchandising department. There should also be representatives from the production, consumer research, product development (if required), engineering, and legal departments. Everyone should have done his homework because the following decisions must be made at this meeting:

- The final design approach must be agreed upon. Matters such as number of colors, copy placement, and the like should be settled so that the designer can create a mock-up of the package or a comprehensive. If more ideas are needed, then more sketches will have to be submitted. However, there should be agreement on the number of colors for purchasing's sake. There is no good reason why this cannot be done.

- Copy should be checked, critiqued, and returned for rewriting if necessary.
- The legal department's representative should review and approve the initial copy.
- The physical package shape should be settled and the preliminary material specifications approved.
- Engineering and production should determine whether any machinery or equipment changes will be needed or whether any production problems will be encountered.
- The consumer research plan, if any, should be approved.
- The product development schedule, if any, should be finalized at this time.
- The project schedule should be updated.

Meeting two will probably be a long one; it must be rigidly controlled and the agenda closely followed.

Preparing for meeting three. The design manager must work closely with the artist to finalize the package design. Copy should be finished. The manager of R&D must see that final specifications are prepared and sent to purchasing for an accurate cost reading. All required tests should be completed. Package purchasing's manager has the major task—to accurately cost out the package components. Sources should be checked for any reproduction problem that might develop with the preliminary design. Production should provide purchasing with the best estimate possible on the number of packages to be produced on the initial run, continue studying any potential problems, and develop product production costs. If its plans have been approved, consumer research implements them. The merchandising manager should be finalizing his thinking on all plans, because after meeting three there will be little, if any, room for change. If required, product development should be working out the last-minute details on the product. Engineering could be quite busy at this time, depending on whether new equipment is needed or not. If new equipment is needed, then it must be located and priced, and lead times must be established.

MEETING THREE

Main Purposes

This is basically a cost-feasibility meeting covering the following matters:

- Packaging costs.
- Production costs.
- Equipment costs (capital expenditures).
- Distribution costs.
- Effect of costs on selling price and markup.
- Overall profit potential.
- Any final-stage designs.

Those who should attend are the managers of design, R&D, package purchasing, and of the involved merchandising department. Representatives should also be present from production, engineering, product development (if needed), consumer research (if needed), and traffic.

Matters that must be attended to at this meeting are:

- Finalize the design so that there will be no more changes.
- Determine whether the packaging costs, based on current design concept and physical specifications, are in line with the profits expected from the product. If costs are out of line, specifications must be trimmed, design reduced, or whatever changes required to reduce the cost made.
- Check the product production costs. Are they within the parameters established? If not what must be done?
- Locate distribution problems if there are any. Solutions must be found now or developed before meeting four.
- Total all the cost factors and compare them with the selling price. Is the project still appealing? If so, the "go" decision is made.

Preparing for meeting four. Preparations by all groups for meeting four mostly deal with cleaning up details. The only man who still has a real project is the manager of design, who must prepare the final design presentation and have everything in place and ready for approval by meeting four.

MEETING FOUR

Main Purposes

- Presentation of final design for approval.
- Submission of final packaging costs.
- Submission of final reports from production and engineering on making and packaging the product.
- Final schedule adjustment.

Those who should attend are the managers of design, R&D, package purchasing, and the involved merchandising department. Representatives should also be present from the production, engineering, consumer research, product development (if needed), traffic, and legal departments.

Decisions that must be made at this meeting are:

- Final approval of design and copy.
- Final approval of any needed equipment.
- Final approval of consumer research and testing schedule.
- Final approval of total cost structure.
- Approval to prepare final artwork.
- Agreement to seek necessary executive approvals on both design and capital expenditures.

After meeting four, the design should be put into final artwork and go to purchasing for production.

The four-meetings system is a mere skeleton or outline waiting to be filled in with a project. It focuses attention where it is needed and, hopefully, makes the job of packaging management in the corporation an easier one. Managing the corporate packaging function may be difficult, but it is an exciting task in a most fascinating, fast-moving, creative, and dynamic industry.